Embedded Control for Mobile Robotic Applications

Embedded Control for Mobile Robotic Applications

Leena Vachhani
Indian Institute of Technology
Bombay

Pranjal Vyas
Advanced Remanufacturing Technology Center
Agency of Science
Technology and Research (A*STAR)
Singapore

Arunkumar G. K.
Indian Institute of Technology
Bombay

IEEE Press Series on Control Systems Theory and Applications
Maria Domenica Di Benedetto, Series Editor

IEEE PRESS

WILEY

Published by John Wiley & Sons, Inc., Hoboken, New Jersey.
Published simultaneously in Canada.

For general information on our other products and services or for technical support, please contact our Customer Care Department within the United States at (800) 762-2974, outside the United States at (317) 572-3993 or fax (317) 572-4002.

Wiley also publishes its books in a variety of electronic formats. Some content that appears in print may not be available in electronic formats. For more information about Wiley products, visit our web site at www.wiley.com.

Library of Congress Cataloging-in-Publication Data applied for:
ISBN: 9781119812388

Cover Design: Wiley
Cover Images: © Thomas Söllner/Getty Images; M-Production/Shutterstock; Bas Nastassia/Shutterstock; Flystock/Shutterstock; Kwanchai Lerttanapunyaporn/Getty Images

Set in 9.5/12.5pt STIXTwoText by Straive, Chennai, India

Contents

Preface

The book aims at the community interested in learning control design for targeting implementation on an embedded platform. The issues in implementing a controller on an embedded platform are usually not considered at the time of designing. This book first gives an overview of implementation issues and then gives its implication on controller performance. The control engineer must ensure that these implications do not disturb the stability of the control system and do not give performance deterioration due to the controller's embedded implementation.

The trend in mobile robotic applications is to use onboard processing and sensing, making the robot self-sufficient. With the objectives of onboard processing and sensing, the controller must guarantee control objectives set for the particular application. While the controller designed using theoretical concepts support the claims on selected control objectives for the robotic application, the faithful implementation of the designed controller is equally important to guarantee the execution of designed controller. Typical issues in embedded implementation are limited memory, limited data width, quantization noise, sampling noise, and limited computational capability. These implementation issues, if ignored, raise the question of stability of the designed controller. Hence, the analysis of controller design is incomplete without considering the issues in implementation. A way to deal with this is to ensure that these embedded implementation issues do not affect stability. The other better way is to consider the issues and limitations of an embedded processor when designing the controller. There is a need of considering the implementation aspects at the time of designing controller using control theoretical concepts.

The book presents the concepts and challenges in designing embedded controller for mobile robotic applications. The approach in this book is to give elementary concepts on embedded designs and use of control concepts for designing efficient embedded controllers for mobile robotic application. The question is now how to use the concepts in future controller designs. The answer lies in the approaches covered in the bottom-up (Chapter 4) and top-down (Chapter 5) approaches. These approaches are presented to provide generic methodologies for embedded controller designs. However, the general methodologies may not be limited to the ones covered in this book. The emphasis in this book is to develop embedded controller designs with a generic approach for catering to variety of control objectives in the larger set of mobile robotic applications. In particular, the book interlaces relevant control theory concepts with embedded design concepts applied to mobile robotic applications. The embedded controller design with a generic approach facilitates straightforward implementation. It is clear that the embedded controller design for mobile robotic applications must cover the topics on embedded design, controller design concepts, and mobile robotic applications. The book presents these concepts in a systematic interlaced manner.

Available **embedded technologies** and the corresponding design efforts to estimate the market value on embedded controller design are introduced in the first chapter. Further, the embedded technologies are mapped to popular embedded processors to date and their requirements and limitations from the controller design perspective.

The **embedded controller** is designed in discrete time as it requires digital implementation, while the operations of a practical system (mobile robot) are in continuous time. The approximate methods that can be used for designing controllers and ensuring stability are discussed in Chapter 2. Furthermore, possible effect of embedded implementation regarding sampling, quantization, and processing-time on stability and performance of controller design are discussed in detail.

The **mobile robotic applications** have a few basic operations or tasks to be implemented. These basic tasks are typically common among many applications. Chapter 3 discusses these common tasks which include 2D and 3D transformations, collision detection and avoidance, localization and navigation. Some of the multi-agent applications are also presented. The perspectives of control design are presented that cover the objectives of these tasks and the feedback collected through the sensors.

Having covered the embedded implementation requirements and objectives of various tasks in mobile robotic applications, two methodologies to design embedded controllers are discussed next. These are *bottom-up* and *top-down*. The **bottom-up methodology** covered in Chapter 4 targets the controller design for a specific embedded platform. In particular, the selection of an embedded platform is followed by the controller design. Alternatively, the **top-down methodology** designs the controller keeping in mind that it is realized on an embedded platform with limited resources. There is no specific way of designing embedded controllers using these methods; hence, the methodologies are illustrated by designing embedded controllers for robotic applications.

The last chapter of the book covers the embedded control design using **Field Programmable Gate Array** (**FPGA**). The FPGA is an embedded platform providing reconfigurability in hardware architecture. Lately, FPGAs have gained popularity, especially in robotic applications, due to their capabilities of parallel processing and multiple Input/Output (I/O) handling. Multiple I/Os do not limit the sensing/actuating dependencies and parallel architecture supports implementation of independent control loops. Basics of FPGA and a methodology to design FPGA architecture for a controller complete the discussion on embedded controller designs.

The target audience for this book are professionals working in embedded control design. The book presents applications of embedded control design to autonomous robotics. Hence, the robotic community would also use the concepts in this book for further control applications in robotics. The book is useful for both academicians and practitioners as it provides content on implementation aspects. The book will also be an excellent resource for designing courses on embedded control and robotics. It covers the discussion on various embedded platforms available in the market, their use in controller implementation, several issues in implementation, stability analysis of designed controller, and two new approaches for designing embedded controllers. Illustrations of these approaches demonstrate the embedded controller designs of typical robotic applications. A chapter on FPGA architecture development for controller design is also a good resource for practitioners. A course on robotics can cover the topics on embedded implementation issues, and controller design approaches discussed in this book.

Leena Vachhani
Mumbai, India
October 2021

Acknowledgments

We acknowledge our mentors, colleagues working in this field, students, friends, and family who have given direct and indirect support in completing this book. It is impossible to call this list complete as it would never be.

This book reflects the work of many past and present students in this field. We would especially like to acknowledge Anindya Harchowdhury, Anupa Sabnis, Aseem V. Borkar, Dhruv Shah, Maria Thomas, Misha Gupta, Mohit Chachada, Mukesh Agarwal, Nipun Agarwal, Nithin Xavier, Sarat Chandra Nagavarapu, Shantanu Thakkar, Shreyas S. G., Siddhesh Wani, Vikranth Reddy Dwaracherla, and Vivek Yogi. Special mention to IIT Bombay faculty working in this field with whom we had several formal and informal discussions. We wish to express our gratitude to Abhishek Gupta, Anirban Guha, Arnab Maity, Arpita Sinha, Bijnan Bandyopadhyay, Hemendra Arya, Kannan Maudgalaya, Ravi Banavar, Sachin Patwardhan, Sukumar Srikant, and P. S. V. Nataraj. Special thanks to K. Sridharan, and B. Ravi for mentoring and guiding Leena in her professional career. Pranjal thanks Leena Vachhani, K. Sridharan, and Sanjeev Kane for their constant support and guidance throughout his career.

We appreciate our most important support system, our friends, parents, and family members. Leena conveys gratitude to Deepak, Deeshan, Dinesh, and Divya for their love, dedication, and consent to sparing her family time on this book. Pranjal is thankful to his wife Hansini for her love, affection, motivation, and constant support during the book-writing phase. He credits his parents' love, blessings, and sacrifice for being in his current professional position.

We would also like to thank the IRCC, IIT Bombay, for providing seed support for initiating research in this field. We thank IIT Bombay for supporting embedded control lab development, DST-SERB, and NRB for their support through research grants.

This book has emerged as a compilation of lecture notes while conducting courses and improving the material based on the received feedbacks. However, when we started compiling our learnings from research and delivering course materials, we realized that it takes efforts more than envisioned to connect the concepts and correct the flow. Lastly, we are thankful to the IEEE-Wiley publishing house for having patience with us and providing feedback on the book proposal.

Acronyms

2D	2 Dimensional
3D	3 Dimensional
ADC	Analog-to-Digital Converter
ALU	Arithmetic and Logic Unit
ASIC	Application-Specific Integrated Circuit
CMR	Car-like Mobile Robot
CORDIC	Coordinate Rotation DIgital Computer
CPLD	Complex Programmable Logic Array
CVM	Curvature Velocity Technique
DAC	Digital-to-Analog Converter
DDMR	Differential-Drive Mobile Robot
DSP	Digital Signal Processor
DWA	Dynamic Window Approach
ECS	Embedded Control System
FPGA	Field Programmable Gate Array
FSM	Finite State Machine
FWSR	Front Wheel Steering Robot
HDL	Hardware Descriptive Language
I/O	Input/Output
IC	Integrated Circuits
IOB	Input–Output Block
LHS	Left-Hand Side
LiDaR	Light Detection and Ranging
LUT	Look Up Table
MPC	Model Predictive Controller
MSB	Most Significant Bit
ND	Nearness Diagram
NRE	Non-Recurrent Engineering
PID	Proportional-Integral-Derivative

PLA	Programmable Logic Array
PLD	Programmable Logic Devices
RHS	Right-Hand Side
RRT	Rapid Exploring Random Trees
SMC	Sliding-Mode Control
VFH	Vector Field Histogram
VLSI	Very Large-Scale Integration
WSN	Wireless Sensor Network
ZoH	Zero-order-Hold

Introduction

An embedded system provides support for sensor interfacing, processing, storing, and/or controlling to facilitate compact and customized solutions. Embedded systems have wide range of applications such as entertainment, communication, security, and automobile. This book focuses on using embedded technology for control system. Targeting the entire control design on an embedded system has many advantages like better performance, portability, optimized area-time ratio, less power requirement, better life, etc. An embedded control system is designed to perform a specific task; therefore, it can optimize on requirements of power, area, and time. Some of the areas where embedded control-systems can replace the traditional control-systems are flight systems, process-control, automobile, and automation industries. Traditional control systems use general-purpose processors. Although the design using traditional control system has flexibility, it cannot be optimized further. The embedded control systems overcome this drawback and can be designed to optimize for area, power, computation time, and accuracy.

There exist many embedded technologies and selecting a right kind of embedded platform from those available in market is another main concern. This selection will affect the performance, power requirement, and life of the control system. Building an embedded control system for a specific application is an intense task. The stand-alone embedded control system has to broadly perform following sub-tasks at every sample instance:

- Process the sensor data (collected as state variables) and extract feedback information from it.
- Compute the control input using a control strategy.
- Issue the commands to the actuators in order to bring the system to next state.

The requirements of the application in terms of its real-time capability, performance in digital world, computation efficiency, etc. have to be studied before choosing the right embedded platform. Techniques for using this embedded platform have to be intensely studied in order to utilize full power of that platform.

There are many issues and challenges to be addressed while implementing a control design on an embedded platform. These issues are as small as digital resolution and as big as real-time requirements. Furthermore, the embedded implementation of control system design must ensure stability and, therefore, there is a need to know simple techniques for guaranteeing that the stability analysis has no effect with embedded implementation. The end product of embedded solution for a control design is an optimized hardware.

The embedded control design for a specific application can target an embedded platform and explore its benefits for controller design and implementation. In this book, we describe the controller designs for mobile robotic applications; hence, following learning objectives and corresponding motive are set addressing a generic approach to design and implement embedded controllers for mobile robotic applications:

- Study of design concepts and popular embedded technologies to know the available benefits and limitations through embedded implementations.
- Understanding existing mobile robot models for representing system for controller designs.
- Learn methods to obtain approximated discrete-time representation for control system. The methodology emphasizes on simple tools to ensure stability analysis is not affected by the embedded implementation.
- Knowledge of a few primitive tasks that are needed for most of the robotic applications helps in forming control objectives.
- The aim of relating the control objectives to the embedded requirements and satisfying the same develops two methodologies. Learning these bottom-up and top-down methodologies using the examples of primitive tasks builds up the techniques for developing embedded controllers for mobile robotic applications.
- For exploiting the benefits of parallel hardware architecture in field programmable gate array (FPGA) in mobile robotic applications, learn a systematic way to implement embedded controller using specific examples.

In order to satisfy the learning objectives, the chapters are organized as follows: Chapter 1 presents the overview of embedded control system followed by kinematic models of 2D and 3D mobile vehicles to develop system model. The design cost involved in embedded controller development is discussed. Various popular commercially available embedded platforms with their limitations and benefits are then addressed. As the controller implemented on an embedded platform performs digital computations, the discrete-time control concepts are covered in Chapter 2. Since the system model (mobile robot) is operating in continuous-time and controller is executed in discrete-time, the corresponding interfaces are discussed and approximate methods that ensure stability are presented. In Chapter 3, the primitive tasks required for most of the mobile robot applications are studied.

Chapters 4 and 5 now describe embedded controller designs for a few primitive tasks and high-level mobile robotic applications covering a generic approach so that controller designs for emerging applications ensure stability and satisfy control objectives. The emerging mobile robotic applications demand for real-time operations with multiple control loops. These requirements can be easily fulfilled by the parallel hardware architecture designs in FPGA. A generic method to design parallel hardware for embedded control is studied in Chapter 6. This book is then summarized in Chapter 7.

About the Companion Website

This book is accompanied by a companion website:

www.wiley.com/go/vachhani/embeddedcontrolforroboticapp

The website include:

- Codes.

1

Embedded Technology for Mobile Robotics

Mobile robot applications are popular in various domains such as security, surveillance, automation, agriculture, and space missions. While the autonomy and capability to perform complex tasks are being possible, it is always required to control the robot operations including low-level control of actuators and processing sensors. It is increasing demand of accommodating all the controls and processing on an embedded system to optimize power consumption, size, and/or cost. The performance of a robotic system not only depends on the attached sensors but also its controller design. Further, the implementation of controller and processing sensors largely contribute to the desired behavior of the robot. Hence, a great deal of importance is to be given for implementation aspects of the controller.

The embedded system solution provides a customized solution for a controller. This customization may target optimization for cost, power, size, speed, or combination of these. Furthermore, the controller on an embedded platform may be designed to achieve real-time requirements. A single platform may be used for multiple controllers working in synchronization. The synchronization or parallel triggering is easily achievable if the architecture for the controllers is designed on a single chip using either Application-Specific Integrated Circuit (ASIC) or Field Programmable Gate Array (FPGA) technology. However, one needs to learn designing architecture to best utilize these technologies.

There exists availability of cross-platforms like *system generator* for high-level programming platforms such as MatLab that makes the job easier for developing embedded controllers using high-level codes. But, unless the control engineer knows about the architectural features of an embedded platform, the advanced properties are difficult to choose. Learning to develop an elementary architecture details would enable a control engineer to optimize the implementation. Hence, the study of Embedded Control Systems (ECSs) involves not only designs of control methodologies but also concepts related to elementary knowledge of embedded requirements. This chapter covers *ECSs*, *Mobile Robots* as a system to be controlled, and *Embedded Technologies* used in existing mobile robots.

Embedded Control for Mobile Robotic Applications, First Edition.
Leena Vachhani, Pranjal Vyas, and Arunkumar G. K.
© 2022 The Institute of Electrical and Electronics Engineers, Inc. Published 2022 by John Wiley & Sons, Inc.
Companion website: www.wiley.com/go/vachhani/embeddedcontrolforroboticapp

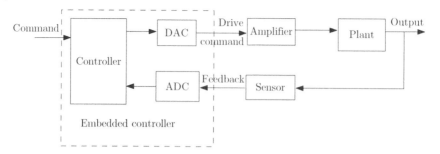

Figure 1.1 Block diagram of an embedded control system

1.1 Embedded Control System

The embedded controller referred here is the controller implemented on an embedded platform. Figure 1.1 illustrates a block diagram of an ECS where the controller is implemented on an embedded platform. The embedded controller is controlling a plant or a system which in our case is a mobile robot. In practice, the plant characteristic is often analog in nature.

The physics of the plant or system is typically modeled using transfer function if the system can be very well described by a linear system. The transfer function of the system describes the input–output relationship in the frequency domain. In particular, a transfer function is the ratio of Laplace transform of output to input of the system. The modern control theory represents a linear system as a state-space model. For representing a system, let the state vector of the system be $X \in \mathbb{R}^n$, input vector to the system be $U \in \mathbb{R}^m$, and output vector be $Y \in \mathbb{R}^p$. The state space representation of the linear system is now given by

$$\dot{X} = AX + BU; \quad Y = CX \tag{1.1}$$

where A is system matrix, B is input matrix, and C is output matrix with appropriate dimensions. For a nonlinear system, the state-space representation is given by

$$\dot{X} = f(X, U); \quad Y = h(X) \tag{1.2}$$

where $f(\cdot)$ and $h(\cdot)$ are nonlinear functions. In an interesting and popular representation of state-space in mobile robotics, the input to system is the derivative of a variable. For example, input to the system is acceleration which is derivative of velocity. The double integrator model is then represented for state variables $X = [X_1 \quad X_2]$ and state-space representation of the system is given by

$$\dot{X}_1 = f(X_1, X_2); \quad \dot{X}_2 = U; \quad y = h(X) \tag{1.3}$$

Similarly, chained form of system description is also very popular and many controllers have been designed for the system in chained form. The input-to-state relation in the chained form of representing the three dimensional system for state vector $X = [x_1 \quad x_2 \quad x_3]^T$ and input vector $U = [u_1 \quad u_2]^T$ is described by

$$\dot{x}_1 = u_1, \quad \dot{x}_2 = u_2, \quad \dot{x}_3 = x_2 u_1 \tag{1.4}$$

In many cases, mobile robot models are represented in double integrator and chained form. Appropriate transformations are used for representing the mobile robot models in these standard forms. An appropriate transformation for control commands accordingly needs to be implemented on the embedded platform. Moreover, the controller design development depends on the application objectives and input–output relationship of the mobile robot represented as a system. While application-specific objectives for embedded implementations are discussed in Chapter 3, the system representation of various popular mobile robots is discussed next.

1.2 Mobile Robotics

There exists variety of mobile robotic platforms that facilitate autonomous and remotely controlled robotic applications. These can be categorized under wheeled robots, aerial vehicles, and underwater vehicles. These vehicles can further have generic categorization of their movements in 2D (planar) and 3D. While these robotic platforms are customized with various payloads like camera, manipulator arm, or application-specific sensors, the common functionality movement can be modeled by describing their kinematic model. The kinematic model provides the relationship between the vehicle velocities and commanded velocities. In this context, the kinematic models describing 2D and 3D motions are as follows.

1.2.1 Robot Model for 2D Motion

To develop robot model, let us consider a point mass moving in a planar environment. The plane is represented in X–Y coordinate frame with a fixed origin O as shown in Figure 1.2. Let the current position of the point representing the robot be

Figure 1.2 A point moving in planar environment

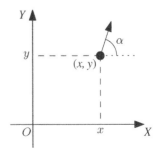

described by its coordinates (x, y) in the fixed X-Y frame. Since the point robot is a moving point, it further needs a description to show its current orientation. Let the current orientation of the point robot be α with respect to the X-axis as shown in Figure 1.2. Now, 2D motion of the point robot is described by $[\dot{x} \ \dot{y} \ \dot{\alpha}]^T$. In order to define the input–output relationship for controller design, the output is the 2D motion of the point robot while the input depends on the command given to the point robot. Accordingly the robot model is developed. Popular robot models and their descriptions are as follows:

1.2.1.1 Generic Model

Let the 2D motion of the robot be given by velocities in the body frame of reference X_b-Y_b, where X_b-Y_b frame is oriented in the forward direction of the robot (at an orientation α) with respect to the inertial frame of reference X-Y. In particular, the velocities are surge-forward v_x and sway-left v_y as shown in Figure 1.3.

Knowing that the rotation by angle α results in transforming a 2D point by matrix R_α, which is given by

$$R_\alpha = \begin{bmatrix} \cos\alpha & -\sin\alpha \\ \sin\alpha & \cos\alpha \end{bmatrix}$$

the kinematic model of generic 2D mobile robot is given by

$$\begin{bmatrix} \dot{x} \\ \dot{y} \end{bmatrix} = R_\alpha \begin{bmatrix} v_x \\ v_y \end{bmatrix} = \begin{bmatrix} \cos\alpha & -\sin\alpha \\ \sin\alpha & \cos\alpha \end{bmatrix} \begin{bmatrix} v_x \\ v_y \end{bmatrix} \tag{1.5}$$

While the generic model describes the commanded velocities in body frame of reference, the commanded velocities in 2D unicycle model are linear and angular velocities as described next.

1.2.1.2 Unicycle Model

Given that the commanded velocities are linear velocity v and angular velocity ω, the model is obtained by relating the angular velocity with change in orientation

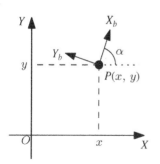

Figure 1.3 Generic 2D robot model

projecting linear velocity v on X and Y axes. These X- and Y-projections define the velocities $\dot{x}$ and $\dot{y}$, respectively. Therefore,

$$\dot{x} = v \cos \alpha$$

$$\dot{y} = v \sin \alpha \tag{1.6}$$

$$\dot{\alpha} = \omega$$

The models of many mobile robots are derived from the unicycle model given by (1.6). Some examples are as follows:

- Differential-Drive Mobile Robot (DDMR).
- Bicycle.
- Car-like Mobile Robot (CMR).

1.2.1.3 Differential-Drive Mobile Robot or DDMR

In order to give the linear and angular motions to the mobile robot, two wheels of the robot in DDMR configuration are independently driven. These independently driven wheels are fixed to the robot's body in the same orientation on a common axis with a center-point M as shown in Figure 1.4. Typically, there is a third caster wheel for balancing the robot which is not driven. Let v_L and v_R be the left and right wheel velocities, respectively. The wheels can move in forward as well as backward directions and, therefore, differential velocity renders the angular motion to the robot. For example, if the left and right wheels are driven by the same speed but in opposite direction, the DDMR rotates at the point M ideally without any linear velocity. Figure 1.5 explains the rotation of the robot in anticlockwise and clockwise directions. The axis of rotation is the center of common axis of two wheels (point M). For the anticlockwise rotation of the DDMR at point M (zero linear velocity), the left wheel is driven in forward direction while right wheel is driven in the reverse direction but with the same speed as that of left wheel as shown in Figure 1.5a. Similarly, Figure 1.5b illustrates the clockwise rotation of the DDMR at point M when the left and right wheels are driven with same speed V in reverse and forward directions, respectively.

Figure 1.4 Differential drive mobile robot schematic illustrating wheel arrangement as seen from top

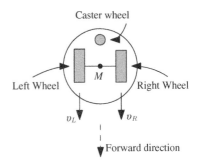

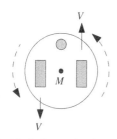

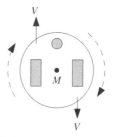

(a) Anticlockwise rotation ($v_L = V$ and $v_R = -V$)

(b) Clockwise rotation ($v_L = -V$ and $v_R = V$)

Figure 1.5 DDMR rotation when wheel velocities are same but in opposite direction

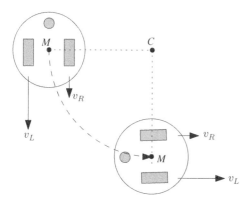

Figure 1.6 Circular motion of the DDMR

In order to further understand the motion of DDMR, let us consider giving differential speeds to two wheels. For example, the left and right wheel velocities are such that $|v_L| > |v_R|$, but in the same direction as shown in Figure 1.6. The DDMR moves in the circular path, center of this circle is point M which is not same as the center of axle joining the wheels.

Now, to formulate the movement of robot's body (linear and angular velocities) in terms of the left and right wheel velocities, we use conversion of angular velocity to linear velocity on the wheels. Let the distance between two wheels be D and the angular rotation of body be with reference to the point M as shown in Figure 1.7. The angular velocity of the robot's body, ω when projected on the left wheel, is $\omega D/2$ as the distance between point M and left wheel is $D/2$. This velocity component is in the forward direction of left wheel, while the direction is reverse when projected on the right wheel. Therefore, total left and right wheel velocities are

$$v_L = v + \frac{\omega D}{2}$$
$$v_R = v - \frac{\omega D}{2}$$

(1.7)

Figure 1.7 Illustration for deriving kinematic model of DDMR

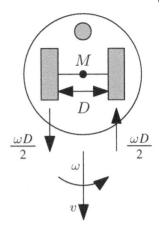

Using (1.7), the linear and angular velocities of DDMR in terms of left and right wheel velocities are given by

$$v = \frac{1}{2}(v_L + v_R)$$

$$\omega = \frac{1}{D}(v_L - v_R)$$

(1.8)

Now, linear and angular velocities obtained in (1.8) and using (1.6), the kinematic model of DDMR is given by

$$\dot{x} = \frac{1}{2}(v_L + v_R)\cos\alpha$$

$$\dot{y} = \frac{1}{2}(v_L + v_R)\sin\alpha$$

$$\dot{\alpha} = \frac{1}{D}(v_L - v_R)$$

(1.9)

The kinematic model as described by (1.9) renders the relationship between the robot's body motion $(\dot{x}, \dot{y}, \dot{\alpha})$ and the commanded velocities (v_L, v_R).

While providing linear motion through wheels of the robot is straightforward and intuitive specially when the wheel orientations are same with respect to the robot body. The differential drive mechanism provides angular motion by independently driving the two wheels. We next discuss the working of driving the robot by front wheel steering.

1.2.1.4 Front Wheel Steering Robot or FWSR

The Front Wheel Steering Robot or FWSR as the name suggests has a wheel or a pair of wheels in front that steers the robot left or right and provides angular motion to the robot. Either the front or rear wheels are driven to provide forward motion. Popular wheel configurations are shown in Figure 1.8. In case of car-like

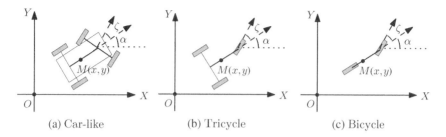

(a) Car-like (b) Tricycle (c) Bicycle

Figure 1.8 Schematics for front wheel steering with various wheel arrangements

(refer Figure 1.8a) robot, the number of wheels is 4 and the front pair of wheels are connected using axle that is steered. Similarly rear pair of wheels are connected using axle so that they are commanded with the same velocity. Similarly, in the Tricycle and Bicycle robots the front wheel is steered. The rear wheels in case of Tricycles again get same actuation. The pair of wheels provides balancing and higher payload capabilities. For simplicity, consider the reference point on the robot's body as any arbitrary point $M(x, y)$ on the line joining mid-points of front and rear wheel axles as shown in Figure 1.9.

Knowing that the pair of wheels either in front or rear are driven in by a same actuation, the kinematics is same for a virtual wheel in the middle of axle of these wheels representing the corresponding pair of wheels. Therefore, the kinematic model is derived using the bicycle configuration with two wheels, one in front and the other in rear. Let the steering angle of front wheel with respect to the body frame of reference be ζ. Similar to the approach for finding kinematic model of the DDMR, the relationship depends on the point of rotation. Figure 1.9 illustrates two cases based on driving wheel. In the first case (refer Figure 1.9a), the driving wheel is the front one, while in the second case (refer Figure 1.9b), the rear wheel is the

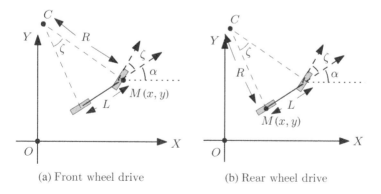

(a) Front wheel drive (b) Rear wheel drive

Figure 1.9 Illustrations for working principle of FWSR

driving one. Let the center of curvature be point C and radius of curvature be R. The reference point $M(x, y)$ is accordingly at the center of front or second wheel. It is clear that the angle made at the point C with the front and rear wheels is ζ.

For the case when the reference point $M(x, y)$ is at the front, we have

$$\sin \zeta = \frac{L}{R}$$

Let the front wheel velocity be v_w. Therefore, angular velocity of the FWSR is

$$\omega = \frac{v_w}{R} = \frac{v_w}{L} \sin \zeta \tag{1.10}$$

Similar to the unicycle model kinematic model, the X and Y projections of velocity vector are now given by

$$\begin{aligned} \dot{x} &= v_w \cos(\alpha + \zeta) \\ \dot{y} &= v_w \sin(\alpha + \zeta) \end{aligned} \tag{1.11}$$

From (1.11) and (1.10), the kinematic model of FWSR with front wheel driven velocity v_w is given by

$$\begin{aligned} \dot{x} &= v_w \cos(\alpha + \zeta) \\ \dot{y} &= v_w \sin(\alpha + \zeta) \\ \dot{\alpha} = \omega &= \frac{v_w}{L} \sin \zeta \end{aligned} \tag{1.12}$$

If the mechanism drives the steering velocity $\dot{\eta}$ as against steering angle η, the kinematic model will include the dynamics of steering angle as well. If the steering wheel velocity is ω_s, it is given by

$$\begin{aligned} \dot{x} &= v_w \cos(\alpha + \zeta) \\ \dot{y} &= v_w \sin(\alpha + \zeta) \\ \dot{\alpha} = \omega &= \frac{v_w}{L} \sin \zeta \\ \dot{\zeta} &= \omega_s \end{aligned} \tag{1.13}$$

Now, let us consider the case of reference point $M(x, y)$ at the rear as shown in Figure 1.9b. In this case, we have

$$\tan \zeta = \frac{L}{R}$$

and therefore, angular velocity of FWSR

$$\omega = \frac{v_w}{R} = \frac{v_w}{L} \tan \zeta \tag{1.14}$$

It is worth noting that the wheel velocity v_w now refers to the rear wheel velocity instead of front wheel velocity in (1.10). Referring to Figure 1.9b and (1.14), we get the kinematic model of rear wheel driven FWSR

$$\dot{x} = v_w \cos \alpha$$
$$\dot{y} = v_w \sin \alpha \qquad (1.15)$$
$$\dot{\theta} = \frac{v_w}{L} \tan \zeta$$

A specific case of $\zeta = \pi/2$ needs attention. We can see that $\dot{\alpha} = \infty$ for $\zeta = \pi/2$. This clearly shows that $\zeta = \pi/2$ case is to be avoided which is precisely the reason for fixing the maximum steering angle less than $\pi/2$ for four wheel automobiles. It is also useful to represent the kinematic models in the form of classical system representations and one of this representation is covered next.

1.2.1.5 Chained form of Unicycle

For the unicycle model described by (1.6), the state vector is $[x \quad y \quad \alpha]^T$ and input vector is $[v \quad \omega]^T$. Therefore, a conversion from unicycle model (1.6) to chained form (1.4) is obtained by a transformation developed in Murray and Sastry (1993) as in

$$x_1 = x, \; x_2 = \tan \alpha, \; x_3 = y$$
$$u_1 = v \cos \alpha, \; u_2 = \omega \sec^2 \alpha$$

Now, taking derivative of each state variable and using (1.6), we get

$$\dot{x}_1 = \dot{x} = v \cos \alpha = u_1$$
$$\dot{x}_2 = \sec^2 \alpha \; \dot{\alpha}$$

Since $\dot{\alpha} = \omega$ from (1.6), we get

$$\dot{x}_2 = \omega \sec^2 \alpha = u_2$$

Similarly,

$$\dot{x}_3 = \dot{y} = v \sin \alpha = v \cos \alpha \tan \alpha$$

But, $x_2 = \tan \alpha$ and $u_1 = v \cos \alpha$, we get

$$\dot{x}_3 = x_2 u_1$$

Similar transformations are applied to convert into other classical forms of system representations. However, it is clear that these transformations require trigonometric functions and sometimes exponential and logarithmic functions to be implemented in the embedded platform.

1.2.1.6 Single Integrator Model of Unicycle

Another common approach in the mobile robot literature (Glotfelter and Egerstedt, 2018) is to use a single integrator model for developing the complex algorithms for single or multi-agent scenarios. This is due to the simple dynamics of the system which allows complex formulation of algorithms and their analysis. Let us understand the aspects behind representing a unicycle model as a single integrator. A single integrator system with states $x(t)$ and input $u(t)$ is given by

$$\dot{x}(t) = u(t) \tag{1.16}$$

Our aim is to explore the relation between the 2D unicycle position variables and the velocity inputs with a pair of single integrator and their corresponding inputs. Let us consider a point $\mathbf{x_d} = (x_d, y_d)$, d distance ahead ($d > 0$) of the position of the unicycle robot at (x, y) on the line along the heading direction. The point can be represented in terms of the state variables of unicycle robot as follows:

$$\begin{bmatrix} x_d \\ y_d \end{bmatrix} = \begin{bmatrix} x \\ y \end{bmatrix} + \begin{bmatrix} d\cos\alpha \\ d\sin\alpha \end{bmatrix} \tag{1.17}$$

The dynamics of the point $\mathbf{x_d}$ corresponding to the dynamics of the robot is now given by

$$\begin{bmatrix} \dot{x}_d \\ \dot{y}_d \end{bmatrix} = \begin{bmatrix} \dot{x} \\ \dot{y} \end{bmatrix} + \begin{bmatrix} -d\sin\alpha\,\dot{\alpha} \\ d\cos\alpha\,\dot{\alpha} \end{bmatrix} \tag{1.18}$$

$$= \begin{bmatrix} v\cos\alpha \\ v\sin\alpha \end{bmatrix} + \begin{bmatrix} -d\omega\sin\alpha \\ d\omega\cos\alpha \end{bmatrix}$$

Rewriting (1.18) as system model renders

$$\dot{\mathbf{x}}_\mathbf{d} = \mathbf{u_d} \tag{1.19}$$

where $\mathbf{u_d} = [u_{xd} \quad u_{yd}]$ and

$$u_{xd} = v\cos\alpha - d\omega\sin\alpha \tag{1.20}$$

$$u_{yd} = v\sin\alpha + d\omega\cos\alpha \tag{1.21}$$

The representation obtained in (1.19) is the single integrator form (refer (1.16)). Now, the inputs to the unicycle robot (v, ω) are transformed using $\mathbf{u_d}$ as follows:

$$\begin{bmatrix} v \\ \omega \end{bmatrix} = \begin{bmatrix} \cos\alpha & -d\sin\alpha \\ \sin\alpha & d\cos\alpha \end{bmatrix}^{-1} \begin{bmatrix} u_{xd} \\ u_{yd} \end{bmatrix} \tag{1.22}$$

The existence of inverse of the matrix used in (1.22) is given in Olfati-Saber (2002).

1.2.1.7 Discrete-time Unicycle Model

The implementation of the controllers on the models described above requires continuous monitoring of the system states and an uninterrupted update of the control command. But in practice, the monitoring and control of the system is done in a discrete manner. The system states are measured every sample time interval and control command is computed and updated in these sampling instances in an ECS as shown in Figure 1.1. Thus, the design of the controllers for a practical system demands a discrete counterpart for the unicycle kinematics presented in (1.6). Let us formalize this by describing the robot states and control commands in discrete time steps. The states and input are assumed to be sampled and calculated every time interval, T_s. Let the states of the robot at nth time instance be $(x(nT_s), y(nT_s), \alpha(nT_s))$. The states at discrete instances are $P[n] = (x[n], y[n], \alpha[n])$. Similarly, let the input to the unicycle robot at nth instant be $(v[n], \omega[n])$. The states of the robot at the successive time instant are denoted by $P[n+1] = (x[n+1], y[n+1], \alpha[n+1])$. A direct approach for discretizing is to use the first-order approximation of Taylor series of the model given in (1.6) or forward difference method. The resultant equations will be as follows:

$$x[n+1] = x[n] + T_s v[n] \cos(\alpha[n])$$
$$y[n+1] = y[n] + T_s v[n] \sin(\alpha[n]) \quad (1.23)$$
$$\alpha[n-1] = \alpha[n] + T_s \omega[n]$$

Now let us analyze this model further to understand the accuracy of this method of discretization. The inputs to the system are assumed to be constant during the time interval $(nT_s, (n+1)T_s]$. Figure 1.10a shows the evolution of the robot for two successive time instances, n and $n+1$, according to the equations given in (1.23). It can be observed that the robot is moving in a straight line for a distance of $v[n]T_s$ in the direction of $\alpha[n]$ and reaches $(x[n+1], y[n+1])$. Also, an instantaneous correction of the orientation is done at $(n+1)$th instant to modify $\alpha[n+1]$. This discretization does not consider the effect of rotation while moving from the location $(x[n], y[n])$ to $(x[n+1], y[n+1])$ and assumes $\omega[n] = 0$. Thus there is an error which accumulates in the difference equation model and may affect the performance of the implementation. Let us consider modifying the difference equation by taking into account the rotation of the unicycle robot during the time interval, $(nT_s, (n+1)T_s]$. The robot will trace an arc when the input to the system, $(v[n], \omega[n])$ is kept constant during the time interval $(nT_s, (n+1)T_s]$. The updated position in the successive time instant is to be calculated considering the circular motion of the robot. The radius of the circle which contains the path of the robot is given by $r[n] = v[n]/\omega[n]$. Referring to Figure 1.10b, the initial position of the robot at nth instant is marked $P[n]$. The circular path taken by the robot during the interval is shown with dash-dotted lines and the final position is marked with $P[n+1]$. The coordinates of the center of the circle $(x_c[n], y_c[n])$ corresponding to the circular path with constant radius $r[n]$ is obtained for nth sample time instance as follows:

$$x_c[n] = x[n] - r[n] \sin(\alpha[n])$$
$$y_c[n] = y[n] - r[n] \cos(\alpha[n]) \quad (1.24)$$

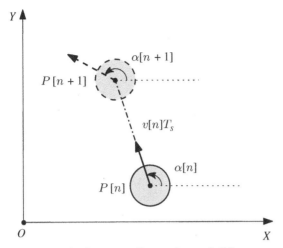

(a) Movement of robot according to forward difference method

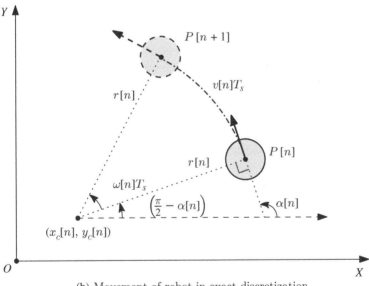

(b) Movement of robot in exact discretization

Figure 1.10   Geometric interpretation of different discretization methods

Further, the coordinates of the robot position at $(n + 1)$th instant are now given by

$$x[n + 1] = x_c[n] + r[n] \sin\left(\alpha[n] + \omega[n]T_s\right)$$
$$y[n + 1] = y_c[n] - r[n] \cos\left(\alpha[n] + \omega[n]T_s\right)$$

(1.25)

On substituting the equations for the coordinates $\left(x_c[n], y_c[n]\right)$ given by (1.24) in (1.25), the discrete equations for unicycle robots are obtained as follows:

$$x[n + 1] = x[n] - r[n] \sin(\alpha[n]) + r[n] \sin\left(\alpha[n] + \omega[n]T_s\right)$$
$$y[n + 1] = y[n] - r[n] \cos(\alpha[n]) - r[n] \cos\left(\alpha[n] + \omega[n]T_s\right)$$
$$\alpha[n - 1] = \alpha[n] + T_s\omega[n]$$

(1.26)

It is worth noting that the equations in (1.26) will reduce to the equations given in (1.23) in the limiting case where $\omega[n] = 0$. These sets of equations are very useful while designing controllers or while using the model for estimating the states as given in Thrun et al. (2005).

1.2.2 Robot Model for 3D Motion

The extension from 2D to 3D is not straightforward due to increase in state variables. While 2D motion can be described by two position coordinates and an orientation, the 3D motion in its most general form is described by three position coordinates and three orientations. The position coordinates can be represented in cartesian (x, y, z), spherical (R, θ, ϕ), or cylindrical (ρ, ϕ, z) as shown in Figure 1.11. Suitable transformations between the coordinate systems are well-known. They are used when the system dynamics require change in representation. In particular, sometimes the problem being addressed can be well represented in spherical coordinates as compared to cartesian or vice versa. Accordingly, the system description is obtained using transformations to convert the position representations. Figure 1.11a also shows body frame of reference X_b-Y_b-Z_b. This body frame of reference is in the direction of inertial frame of reference $(X-Y-Z)$. In particular, the vehicle (or 3D rigid body) located at a point $P(x, y, z)$ in Figure 1.11a has orientation same as that of inertial. The unit directions in the corresponding inertial frame of direction for spherical and cylindrical coordinate system are shown as e_r-e_θ-e_ϕ and e_ρ-e_θ-e_z, respectively.

The orientation of the robot in 3D (or the orientation of the body frame) with respect to an inertial frame can be represented using three composed rotations. The angles of rotations are called Euler angles. There exist many representations for orientation in 3D using Euler angles depending on the choice of the axis of rotation such as *XYZ*, *ZYX*, *ZYZ*, and *ZXZ*. The axis of rotations can be of inertial frame or of the body frame. For example, $Z_b Y_b X_b$ Euler angles also known as Yaw–Pitch–Roll rotations or Tait–Bryan angles are most popular in 3D vehicle descriptions. Henceforth we follow Yaw–Pitch–Roll convention of angles to

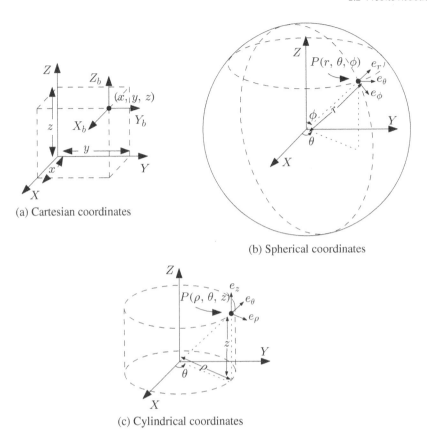

(a) Cartesian coordinates

(b) Spherical coordinates

(c) Cylindrical coordinates

Figure 1.11 Typical 3D position representations

represent the orientation. A generic rotation of body frame in 3D that can be represented using Euler angles which are yaw, pitch, and roll is shown in Figure 1.12. Refer to the rotation of body frame of reference in Figure 1.12. The rotation about Z_b-axis is the yaw (α), while rotations about Y_b-axis and X_b-axis are pitch (β) and roll (γ) respectively. Also, note that the order of the rotations is important and not interchangeable. The transformation matrix for each of the rotations from inertial frame to body frame is defined by elementary rotations along each of the axes. Following are the transformation matrices corresponding to yaw, pitch, and roll with respect to Z_b, Y_b, and X_b axes, respectively, by yaw angle α, pitch angle β, and roll angle γ.

$$\mathcal{R}_{Z_b,\alpha} = \begin{bmatrix} \cos \alpha & \sin \alpha & 0 \\ -\sin \alpha & \cos \alpha & 0 \\ 0 & 0 & 1 \end{bmatrix} \tag{1.27}$$

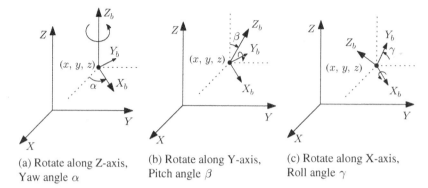

(a) Rotate along Z-axis,
Yaw angle α

(b) Rotate along Y-axis,
Pitch angle β

(c) Rotate along X-axis,
Roll angle γ

Figure 1.12 Illustration of 3D orientations

$$\mathcal{R}_{Y_b,\beta} = \begin{bmatrix} \cos\beta & 0 & -\sin\beta \\ 0 & 1 & 0 \\ \sin\beta & 0 & \cos\beta \end{bmatrix} \tag{1.28}$$

$$\mathcal{R}_{X_b,\gamma} = \begin{bmatrix} 1 & 0 & 0 \\ 0 & \cos\gamma & \sin\gamma \\ 0 & -\sin\gamma & \cos\gamma \end{bmatrix} \tag{1.29}$$

Since these transformations are rotation matrices, they are orthogonal and satisfy the following:

$$\mathcal{R}\mathcal{R}^T = \mathcal{R}^T\mathcal{R} = I; \quad det\,\mathcal{R} = 1 \implies \mathcal{R}^{-1} = \mathcal{R}^T \tag{1.30}$$

Now, if the surge, sway, and roll velocities in body frame are v_x, v_y, and v_z, respectively, then

$$\begin{bmatrix} \dot{x} \\ \dot{y} \\ \dot{z} \end{bmatrix} = \left(\mathcal{R}_{X_b,\gamma} \mathcal{R}_{Y_b,\beta} \mathcal{R}_{Z_b,\alpha} \right)^{-1} \begin{bmatrix} v_x \\ v_y \\ v_z \end{bmatrix} \tag{1.31}$$

$$= \mathcal{R}_{Z_b,\alpha}^{-1} \mathcal{R}_{Y_b,\beta}^{-1} \mathcal{R}_{X_b,\gamma}^{-1} \begin{bmatrix} v_x \\ v_y \\ v_z \end{bmatrix}$$

The transformation matrix $\mathcal{R}_{Z_b,\alpha}^{-1} \mathcal{R}_{Y_b,\beta}^{-1} \mathcal{R}_{X_b,\gamma}^{-1}$ is calculated as follows:

$$\mathcal{R}_{Z_b,\alpha}^{-1} \mathcal{R}_{Y_b,\beta}^{-1} \mathcal{R}_{X_b,\gamma}^{-1} = \begin{bmatrix} \cos\alpha & -\sin\alpha & 0 \\ \sin\alpha & \cos\alpha & 0 \\ 0 & 0 & 1 \end{bmatrix} \begin{bmatrix} \cos\beta & 0 & \sin\beta \\ 0 & 1 & 0 \\ -\sin\beta & 0 & \cos\beta \end{bmatrix} \begin{bmatrix} 1 & 0 & 0 \\ 0 & \cos\gamma & -\sin\gamma \\ 0 & \sin\gamma & \cos\gamma \end{bmatrix}$$

$$= \begin{bmatrix} c_\alpha c_\beta & c_\alpha s_\beta s_\gamma - s_\alpha c_\gamma & c_\alpha s_\beta c_\gamma + s_\alpha s_\gamma \\ s_\alpha c_\beta & s_\alpha s_\beta s_\gamma + c_\alpha c_\gamma & s_\alpha s_\beta c_\gamma - c_\alpha s_\gamma \\ -s_\beta & c_\beta s_\gamma & c_\beta c_\gamma \end{bmatrix} \tag{1.32}$$

where s_* and c_* represent $\sin(*)$ and $\cos(*)$, respectively. In order to map the 3D angular velocities of the vehicle ω_x, ω_y, and ω_z with the time derivatives of yaw–pitch–roll $\dot{\alpha}$, $\dot{\beta}$, and $\dot{\gamma}$, we consider a map $E(\alpha, \beta, \gamma)$ that maps 3D angular velocities to time derivatives of roll-pitch–yaw. In particular, we consider

$$\begin{bmatrix} \dot{\gamma} \\ \dot{\beta} \\ \dot{\alpha} \end{bmatrix} = E(\alpha, \beta, \gamma) \begin{bmatrix} \omega_x \\ \omega_y \\ \omega_z \end{bmatrix}$$

The Euler angle rates are related to the 3D angular velocities of the vehicle in the body frame under the following transformation.

$$\begin{bmatrix} \omega_x \\ \omega_y \\ \omega_z \end{bmatrix} = \begin{bmatrix} \dot{\gamma} \\ 0 \\ 0 \end{bmatrix} + T_{X_b,\gamma} \begin{bmatrix} 0 \\ \dot{\beta} \\ 0 \end{bmatrix} + T_{X_b,\gamma} T_{Y_b,\beta} \begin{bmatrix} 0 \\ 0 \\ \dot{\alpha} \end{bmatrix} \tag{1.33}$$

$$= \begin{bmatrix} 1 & 0 & -\sin\beta \\ 0 & \cos\gamma & \cos\beta\sin\gamma \\ 0 & -\sin\gamma & \cos\beta\cos\gamma \end{bmatrix} \begin{bmatrix} \dot{\gamma} \\ \dot{\beta} \\ \dot{\alpha} \end{bmatrix}$$

$$= E^{-1}(\alpha, \beta, \gamma) \begin{bmatrix} \dot{\gamma} \\ \dot{\beta} \\ \dot{\alpha} \end{bmatrix}$$

Further E is computed from (1.33) as follows.

$$E(\alpha, \beta, \gamma) = \begin{bmatrix} 1 & \sin\gamma\tan\beta & \cos\gamma\tan\beta \\ 0 & \cos\gamma & -\sin\gamma \\ 0 & \sin\gamma/\cos\beta & \cos\gamma/\cos\beta \end{bmatrix} \tag{1.34}$$

In summary, the kinematic model of the 3D vehicular motion is given by

$$\begin{bmatrix} \dot{x} \\ \dot{y} \\ \dot{z} \\ \dot{\gamma} \\ \dot{\beta} \\ \dot{\alpha} \end{bmatrix} = \begin{bmatrix} c_\alpha c_\beta & c_\alpha s_\beta s_\gamma - s_\alpha c_\gamma & c_\alpha s_\beta c_\gamma + s_\alpha s_\gamma & 0 & 0 & 0 \\ s_\alpha c_\beta & s_\alpha s_\beta s_\gamma + c_\alpha c_\gamma & s_\alpha s_\beta c_\gamma - c_\alpha s_\gamma & 0 & 0 & 0 \\ -s_\beta & c_\beta s_\gamma & c_\beta c_\gamma & 0 & 0 & 0 \\ 0 & 0 & 0 & 1 & s_\gamma t_\beta & c_\gamma t_\beta \\ 0 & 0 & 0 & 0 & c_\gamma & -s_\gamma \\ 0 & 0 & 0 & 0 & s_\gamma/c_\beta & c_\gamma/c_\beta \end{bmatrix} \begin{bmatrix} v_x \\ v_y \\ v_z \\ \omega_x \\ \omega_y \\ \omega_z \end{bmatrix} \tag{1.35}$$

Now, to generate the linear and angular velocities of a 3D vehicle, there exist various configurations. Two of these configurations are discussed next. The first configuration covers an aerial vehicle design, while the second configuration covers an underwater vehicle.

1.2.2.1 Quadcopter – An Aerial Vehicle

The quadcopter as the name suggests has four propellers mounted on the corners of a virtual square.
The diagonals of this virtual square are connected as
a base. A rough schematic of a typical quadcopter is
shown in Figure 1.13. Four propellers are placed in
front, right, rear, and left. Let the angular speeds of
front, right, rear, and left propellers be Ω_1, Ω_2, Ω_3,
and Ω_4, respectively. The thrust generated through
a propeller is proportional to square of the corresponding angular speed. Hence, the upward thrust
(in Z_b direction) of the quadcopter τ_z is given by

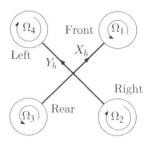

Figure 1.13 Schematic of a
quadcopter

$$\tau_z = k_1(\Omega_1^2 + \Omega_2^2 + \Omega_3^2 + \Omega_4^2) \tag{1.36}$$

where k_1 is a proportionality constant with appropriate dimensions. Now, the
thrust in yaw (angular orientation around Z_b axis) is due to the differential thrust
generated by left and right propellers, or the thrust in yaw direction is given by

$$\tau_\alpha = k_2(-\Omega_2^2 + \Omega_4^2) \tag{1.37}$$

Similarly, thrust generated in pitch and roll directions is given by

$$\tau_\beta = k_3(-\Omega_1^2 + \Omega_3^2) \tag{1.38}$$

$$\tau_\gamma = k_4(\Omega_2^2 + \Omega_4^2 - \Omega_1^2 - \Omega_3^2) \tag{1.39}$$

Note that the proportionality constants k_2, k_3, and k_4 have appropriate dimensions.

Now, if the state vector of a 3D vehicle is given by $X = [x\ y\ z\ \alpha\ \beta\ \gamma]^T$. It
is clear from (1.36)–(1.39), the actuation (input vector) to the quadcopter is
$[\tau_z\ \tau_\alpha\ \tau_\beta\ \tau_\gamma]^T$. We know that the thrust is proportional to the acceleration.
Moreover, the vertical thrust τ_z is responsible for providing accelerations in
X_b–Y_b–Z_b directions depending on the instantaneous pitch β and roll γ. Therefore, the relation of thrusts with corresponding accelerations and appropriate
proportionality constants is described by

$$\begin{bmatrix} \tau_z & \tau_\alpha & \tau_\beta & \tau_\gamma \end{bmatrix}^T = \begin{bmatrix} k_1'g(\dot{v}_x, \dot{v}_y, \dot{v}_z) & k_2'\dot{\omega}_\alpha & k_3'\dot{\omega}_\beta & k_4'\dot{\omega}_\gamma \end{bmatrix}^T$$

This way the quadcopter model is represented as a double integrator model as
described by (1.3).

1.2.2.2 Six-Thrusters Configuration – An Underwater Vehicle

As the name suggests, the vehicle has six thrusters to provide control on surge
(X_b), sway (Y_b), heave (Z_b), yaw (α), and pitch (β). The underwater world is upside
down; hence, the body frame of reference (X_b–Y_b–Z_b) shows Z_b in the downward
direction. The configuration in which they are placed, provides the capability for

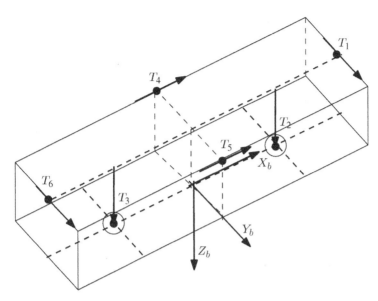

Figure 1.14 Schematic of six-thrusters configuration; $T_1 - T_6$ show typical placement and directions of six thrusters

such a control. A schematic of six-thrusters configuration is shown in Figure 1.14. The six thrusters shown using T_1 to T_6 present a typical positioning and direction of corresponding thrusters. As in differential drive vehicle, two actuators (motors) provide the forward and yaw motion, the thrusters T_4 and T_5 provide surge and yaw motions to the 3D underwater vehicle. Likewise, T_1 and T_6 provide sway and yaw motions, while T_2 and T_3 provide heave and pitch motions. Therefore, if the corresponding actuation velocities of thrusters are $\Omega_1 - \Omega_6$, the thrusts $\tau_x, \tau_y, \tau_z, \tau_\alpha$, and τ_β in X_b, Y_b, Z_b, α, and β directions are given by

$$
\begin{aligned}
\tau_x &= k_1(\Omega_4^2 + \Omega_5^2) \\
\tau_y &= k_2(\Omega_1^2 + \Omega_6^2) \\
\tau_z &= k_3(\Omega_2^2 + \Omega_3^2) \\
\tau_\alpha &= k_4(\Omega_4^2 - \Omega_5^2) + (\Omega_1^2 - \Omega_6^2) \\
\tau_\beta &= k_5(\Omega_3^2 - \Omega_2^2)
\end{aligned}
\tag{1.40}
$$

The proportionality constants $k_1 - k_5$ are appropriately dimensioned and used here to show the relationship between the thrusters' actuation and thrusts generated in various directions.

With this background of various models for 2D and 3D mobile robots, it is important to understand the perspective of emerging embedded technology. The next section presents various aspects of latest embedded technology.

1.3   Embedded Technology

An embedded controller computes a dedicated function, with real-time computing constraints. The controller output can also be generated using general-purpose computing machines which may overcome real-time computing constraints as well; however, the undesirable architecture features and resources consume considerable power. This power consumption is at least 20 times more than the power required for computing dedicated function. The embedded platforms provide a cost-effective solution with low-power consumption and size. It also facilitates easy installation. Typically, an embedded controller is designed to provide a customized solution for performing a specific task and its objective is to consume just sufficient embedded resources according to its needs. In particular, while designing the embedded controllers, the objective is not to use extra peripheral/resource other than the ones available on the selected embedded platform. A holistic approach of application-oriented controller implementation has a major challenge in designing the embedded controller maintaining the trade-off between the limitations of embedded computing and combined objectives to satisfy size, power, cost, cooling, and weight requirements. The selection of embedded platform decides the constraints on developing controller. The controller may also need to operate in extreme operating conditions such as vibration, shock, extreme heat and cold, or radiation. These operating conditions need to be considered while selecting an embedded platform.

The challenge in designing embedded controller is to optimize the design metrics. Following are a few common design parameters:

- *NRE (Non-Recurrent Engineering) cost*: A one-time design cost involves design iterations. The design may undergo various iterations where embedded controller design and its interface may require redesigning hardware as well as software.
- *Unit cost*: This refers to the per unit cost after designing.
- *Size*: The spatial area occupied by the controller.
- *Performance*: This is mainly weighed in terms of the response time (latency) and throughput of the controller. The latency is defined as the time required for sensing, executing the controller functionality, and actuating. Throughput is defined as the number of controller outputs generated per unit time. Therefore, latency and throughput are reciprocal of each other if a single processor without pipelining is being used. For example, if the latency of a controller is 0.1 second, then the controller output is generated 10 times per second or throughput of the controller is 10 control outputs/s.
- *Power consumption*: The battery life and the cooling requirements are dictated by the total power consumption of the design.

- *Flexibility*: This refers to the ability to change functionality without incurring heavy NRE cost.
- *Time-to-market*: This includes time to design, test, and manufacture.
- *Safety*: The design must be able to operate in the specified operating conditions.
- *Correctness*: This reflects confidence in the design.

In a typical scenario, a trade-off between power, size, performance, and NRE cost is required to be achieved. The effort in reducing the power consumption usually increases the size, performance, and subsequently, NRE cost. An optimal design for embedded controller is achieved by understanding different technologies available in embedded platforms. An in-depth understanding is not required, but a fair idea of these technologies helps in selecting an appropriate embedded platform. A comprehensive idea of these technologies is covered from the perspectives of processor and Integrated Circuit (IC) technologies. These technologies cover commercially available embedded platforms used in mobile robots like Microcontroller, FPGA, Digital Signal Processor (DSP), etc.

1.3.1 Processor Technology

In this technology, architecture of the selected processor dictates the functionality of the controller. This can be classified in following three categories:

1. *General-purpose processor*: The controller functionality is implemented through software. The evaluation of design parameters is as follows:
   - The NRE cost is low (desirable feature)
   - The design is flexible (desirable feature)
   - Power consumption is high (undesirable feature)
   - Occupies a larger area (undesirable feature).
2. *Single-purpose processor*: The processor is designed to generate a controller output and results in a system-on-chip solution. The evaluation of design parameters is as follows:
   - It achieves low latency or high throughput (desirable feature)
   - It occupies smaller area and consumes low power (desirable feature)
   - It typically achieves low unit cost (desirable feature)
   - The design takes longer time to optimize; therefore, NRE cost is usually high (undesirable feature)
   - The design is not flexible (undesirable feature).
3. *Application-specific processor*: These processors are designed for solving a class of problems, e.g.; embedded control, image processing, and telecommunication. Microcontroller technology is most appropriate for designing embedded controllers. The evaluation of design parameters is as follows:
   - It achieves considerably low latency or high throughput

- It occupies smaller area and consumes low power
- It typically achieves low unit cost
- The design does not take longer time to optimize; therefore, NRE cost is moderate
- The design is not completely flexible, but has a flexibility to modify the controller parameters and functionality by reprogramming the processor.

1.3.2 IC Technology

The ICs technology is independent of the processor technology. This technology refers to layers created over semiconductor and designing appropriate masks for creating these layers. These sets of masks define the layout of the design. Figure 1.15 shows different layers. The bottom-most is the transistor layer. A set of transistors connected in a predefined manner provides the functionality of different logic components. This layer is referred to as *logic component* layer. The interconnections of logic components form the *wire* layer. The design is complete once the *wire* layer is designed. Classification of the embedded system is also based on these layer formations. The classification is as follows:

1. *Full custom/VLSI* (*Very Large-Scale Integration*): This technology optimizes all the layers. In particular, the system is designed at the transistor level. The evaluation of design parameters is as follows:
   - It requires longer designing time resulting in high NRE cost
   - It results in small size and low power requirements.
   This technology is usually preferred in extremely performance-critical conditions with high production requirement.
2. *Semiconductor ASIC (Application-Specific Integrated Circuit)*: The system is designed at middle layer of logic components. The transistor layer is already built and fixed. The evaluation of design metric is as follows:
   - It achieves considerably good performance
   - It results in small size and lesser NRE cost as compared to VLSI technology.
3. *Programmable Logic Devices* (*PLDs*): The transistor and logic connector level layers are fixed in this technology. It creates or destroys connections between the wires that connect gates by programming the PLD. The popular PLDs in

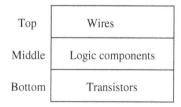

Top	Wires
Middle	Logic components
Bottom	Transistors

Figure 1.15 Various layers of IC technology

market are PLA (Programmable Logic Array), FPGA, and CPLD(Complex Programmable Logic Array). The evaluation of design parameters is as follows:

- It results in low NRE cost, but high unit cost
- It results in bigger size and higher power consumption than that achieved by the ASIC technology
- It achieves reasonable performance.

1.4 Commercially Available Embedded Processors

There are many popular commercially available embedded processors based on the processor and IC technologies that are being used in the embedded control designs. A brief idea on these commercial embedded platforms is needed prior to designing the embedded controllers. Learning the features of the architecture would provide understanding of the capabilities and limitations that is useful in selecting the embedded platform for the controller design and implementation. The categorization of processor with their commercial names is presented next.

1.4.1 Microprocessor

The general-purpose technology used for microprocessor includes basic architecture elements like arithmetic and Logic Unit (ALU), internal registers and memory and I/O (Input/Output) port interfaces. The processor is selected based on the computation power and interface capabilities with peripherals in terms of its data and address lines. For example, a 16-bit microprocessor like 8086 has 16-bit internal registers, 16-bit ALU, and 16-bit external data bus, but 20-bit address lines. The latest processors like Intel i7 also has multiple cores for creating parallel threads, and complex computations for optimal and nonlinear controls can then be possible with real-time performance. However, the general-purpose processor lacks customization capabilities and typically is not very suitable for embedded control designs. Many popular processors have been used in control designs. Popular choices in this category are Intel Core i7, AMD Athlon X7, Intel 8085, Intel 8086, ARM NEON, Zilog Z80, etc.

The latest category of microprocessors provide customized solutions through embedded Linux. The Linux kernel can be customized for the controller applications and thus can be light-weight. Available processors in the market under this category are ARM Cortex series, XScale PXA series, MIPS MSP series, AT91 by ATmel, PowerPC, Freescale MPC5200, etc. This choice of embedded platform facilitates easier interfacing and real-time performance, but requires specialized coding skills.

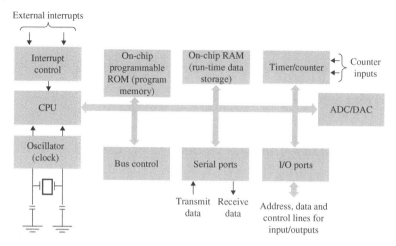

Figure 1.16 Block diagram of a typical microcontroller

1.4.2 Microcontroller

Microcontroller provides a customized solution as compared to the microprocessor that requires external peripherals like memory, I/O ports, timers, etc. A typical microcontroller has a CPU with program and data memory, I/O port interfaces, ADC/DAC, timer(s) that can be used as counter(s) and interrupt handling unit. A block diagram of a typical microcontroller is shown in Figure 1.16. With a few peripherals on chip, the microcontroller can be a standalone solution for controller implementations in many applications. The program memory can support the control-law, and ADC/DAC with I/O ports can provide interface to actuators and sensors. The computation capabilities are limited by the sizes of data registers. For example, an 8-bit microcontroller can perform 8-bit operations efficiently, but the performance degrades with computations requiring complex requiring floating-point operations. Similarly, program memory limits the code length corresponding to the control-law implementation.

Popular microcontrollers in the market are Altera Nios-16 bit, Intel 8051, Atmel AT89 series (Intel 8051 architecture), Atmel ATxmega series (AVR architecture), Cypress PSoC series (ARM Cortex), Freescale (ARM architecture), etc.

1.4.3 Field Programmable Gate Arrays (FPGA)

The FPGAs support system-on-chip design by facilitating hardware reconfigurable facility. It is easy to understand the reconfigurability of FPGAs by learning its architecture features. Figure 1.17 shows basic building blocks of FPGA

Figure 1.17 Typical architecture layout of an FPGA; LUT stands for look-up table and IOB stands for input–output block

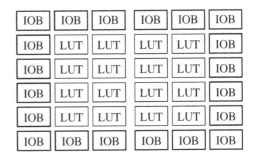

architecture. Basic building blocks are Look-Up Tables (LUTs) at the center and Input–Output Blocks (IOBs) at the periphery. The LUTs are programmable to contain logic. The IOBs interface with sensors/actuators and programmable to act as input or output or both. Each FPGA has many LUTs and IOBs. For example, the smallest FPGA chip in Xilinx Spartan 3E family, XC3S100E has 2160 logic cells and 108 user IO pins. The programmable LUTs and IOBs provide support for hardware reconfigurability of FPGAs. The FPGAs have been widely used in designing hardware accelerators by exploiting the parallel processing in hardware. Any hardware module (programmed through LUTs) can be copied multiple times to create parallel functionality in FPGA for real-time performance. Likewise, parallel processing of multiple sensors and actuators (may not be of same kind) is easily reconfigured in FPGA avoiding time delays in servicing events generated by respective input and/or output device.

Popular choices are Xilinx Spartan series, Xilinx Vertex series, Altera Cyclone-V, Altera Stratix-V, etc. For the controller design perspective, FPGA is a good choice when there is a need of (i) real-time applications, (ii) I/O requirement is high, and/or (iii) taking benefits of parallel processing.

1.4.4 Digital Signal Processor

The DSPs are specifically designed for the needs of signal processing using specialized microprocessor architecture. The architecture comprises large data paths and memory with facility of matrix manipulations. The DSPs in control systems are typically used to deal with large amount of sensor data and specialized needs on processing the data to provide tangible feedback to the controller.

A few popular DSPs are Texas Instruments TMS320 series and C6000, Blackfin family, Embedded general purpose processor like OMAP3 includes ARM Cortex-A8 and C6000. The high-level programming environments are available that use "C-language" like constructs for designing algorithms using DSP.

1.5 Notes and Further Readings

The basic concepts on mobile robotics covered here are to connect with the control theory. Various kinematic models used with 2D and 3D mobile robots are expressed as system representations. The suggested reading on the detailed kinematic and dynamic models of mobile robots is Siciliano and Khatib (2016) or Fossen (1994). One may refer to Olfati-Saber (2002) for further reading on unicycle model to single integrator. The reader is encouraged to develop system representations defining states, inputs, outputs, and relations using dynamic and kinematic models. For the embedded controller design, the basics of embedded processor technology are presented for a naive person to select the embedded processor from the design perspective. Suggestions for further reading on embedded systems are Givargis and Vahid (2012) and Wescott (2006).

There has been a growing interest in learning embedded control concepts. Demonstrating the embedded control concept with application benefits has always been an effective course delivery method. The concepts developed in this chapter can supplement developing course modules on "Embedded Control for Mobile Robotics." There are many concepts on developing embedded control laboratory for giving hands-on experience on embedded platforms like microcontroller (Moallem, 2004) and Real-Time Embedded Linux (Martí et al., 2010). The control education with a low-cost solution using real-time execution of control-law (Krauss and Croxell, 2012) uses PC for the calculations and the interfaced microcontroller for the execution of control law in real time and analog-to-digital/digital-to-analog conversions avoiding the need of real-time operating system.

Any embedded implementation is in discrete-time or event-driven. The computations are based on the system clock and require designing of digital logic. Once the digital logic is designed, the effects of quantization and processing delay in sequential logic account for the performance of control loop. While the accuracy and resolution of computations depend on the quantization, the control loop time depends on the processing and sampling times. Therefore, these effects limit the faithful implementation of designed controller.

As explained in Section 1.1, the controller is implemented on an embedded platform while the system is typically continuous-time in nature. The controller design in discrete-time is covered next to understand the interfacing with the analog world and embedded implementation.

2

Discrete-time Controller Design

In the Chapter 1, a few popular system models used in mobile robotics were explained. These system models represent continuous-time system. However, when the controller is to be implemented on an embedded platform, the controller design is in digital domain. Therefore, this chapter covers the discrete-time controller design methods and its interfacing with the continuous-time mobile robot model. Furthermore, the aspects covering the issues in digital (embedded) implementation of discrete-time controller are discussed.

Refer to Figure 2.1 to understand the interfacing of embedded controller with a real-world system or plant (a mobile robot in our case). The plant operates in continuous-time, and operations are analog in nature. Therefore, interfacing analog plant with embedded controller needs Analog-to-Digital conversion (Sample and Hold) and Digital-to-Analog conversion (Zero-order Hold [ZoH]).

The input and output signals to the plant are continuous-time, and their frequency domain representations are obtained using the Laplace transform. The corresponding transfer function is expressed in complex angular frequency s. While the input and output signals to the embedded controller are discrete-time and Z-transform is used to obtain their frequency domain representations. Hence, the controller transfer function is expressed in complex frequency z. We now describe various approximate relationships between Laplace transform and Z–transform in Section 2.1. The objective is to use approximate technique to obtain the transfer function for discrete-time equivalent model of the analog plant.

2.1 Transfer Function for Equivalent Discrete-time System

The discrete-time equivalent of a continuous-time signal is obtained by sampling the signal. Let $x(t)$ be a continuous-time signal. The one-sided Laplace transform

Embedded Control for Mobile Robotic Applications, First Edition.
Leena Vachhani, Pranjal Vyas, and Arunkumar G. K.
© 2022 The Institute of Electrical and Electronics Engineers, Inc. Published 2022 by John Wiley & Sons, Inc.
Companion website: www.wiley.com/go/vachhani/embeddedcontrolforroboticapp

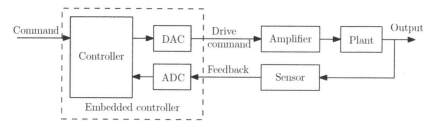

Figure 2.1 Block diagram of an embedded control system

$X(s)$ of signal $x(t)$ is given by

$$X(s) \triangleq \int_0^\infty x(t)e^{-st}dt$$

Let $x_q(t)$ be the continuous-time representation of sampled sequence $x(nT_s)$ or $x[n]$, where T_s is the sampling period. Thus, signal $x_q(t)$ is given by

$$x_q(t) = \sum_{n=0}^\infty x[n]\delta(t - nT_s)$$

Therefore,

$$X_q(s) = \int_{0-}^\infty x_q(t)e^{-st}dt$$

$$= \int_{0-}^\infty \sum_{n=0}^\infty x[n]\delta(t - nT_s)e^{-st}dt$$

$$= \sum_{n=0}^\infty x[n] \int_{0-}^\infty \delta(t - nT_s)e^{-st}dt$$

$$= \sum_{n=0}^\infty x[n]e^{-nsT_s}$$

The unilateral or one-sided Z-transform is obtained by substituting $z = e^{sT_s}$,

$$X(z) = \sum_{n=0}^\infty x[n]z^{-n}$$

But the relationship $z = e^{sT_s}$ renders an inconvenient method for converting Laplace transform of a signal to its Z-transform. Our aim is to find an approximate method for this conversion. The equivalent difference operation of differentiation is considered for this purpose. Let us investigate the approximation using different methods.

1. *Backward difference method*: The approximation is obtained by considering previous sample value and given by

$$\frac{dx}{dt} \approx \frac{x[n] - x[n-1]}{T_s}$$

$$\text{or } sX(s) \approx \frac{X(z) - z^{-1}X(z)}{T_s}$$

$$\text{or } s \approx \frac{1 - z^{-1}}{T_s} \tag{2.1}$$

Note that the approximation is made at various levels, viz. the Laplace transform of the signal $X(s)$ is considered equivalent to the Z-transform of the sampled sequences $X(z)$. Equation (2.1) suggests that a pole at $s = a$ would result in a pole at $z = 1/(1 - aT_s)$. Therefore, a stable pole ($a < 0$) in continuous domain representation of a system results in a stable pole in its equivalent discrete domain representation using this approximation. But an unstable pole ($a > 0$) in s-domain representation of system results in a pole at $1/(1 - aT_s)$ in its equivalent discrete-time representation. This would be a stable pole if $|1/(1 - aT_s)| < 1$ or $|(1 - aT_s)| < 1$. In particular, an unstable pole ($a > 0$) in continuous domain is approximated to a stable pole in discrete domain if the sampling time $T_s > 2/a$. Clearly, this approximation is a valid approximation if the constraint on sampling time $T_s < 2/a$ is satisfied. This ensures unstable pole in continuous domain of a system representation, maps to an unstable pole in its discrete domain representation.

2. *Forward difference method*: The approximation is obtained by considering next sample value and given by

$$\frac{dx}{dt} \approx \frac{x[n+1] - x[n]}{T_s}$$

$$\text{or } sX(s) \approx \frac{zX(z) - X(z)}{T_s}$$

$$\text{or } s \approx \frac{z - 1}{T_s} \tag{2.2}$$

Equation (2.2) suggests that a pole at $s = a$ would result in a pole at $z = (1 + aT_s)$. Therefore, an unstable pole ($a > 0$) in continuous-time representation of a system results in an unstable pole in its discrete-time representation using this approximation. But a stable pole ($a < 0$) in continuous-time system representation results in a pole at $(1 - |a|T_s)$ in its discrete-time representation. This would be an unstable pole if $|1 + aT_s| > 1$. In particular, a stable pole in the continuous-time representation of a system is mapped to an unstable pole in its discrete-time representation if the sampling time $T_s > \frac{2}{a}$. Clearly, this approximation also results in the same constraint as that obtained with backward difference approximation.

3. *Center difference method*: The approximation is obtained by averaging the previous and next sample values and given by

$$\frac{dx}{dt} \approx \frac{x[n+1] - x[n-1]}{2T_s}$$

$$\text{or } sX(s) \approx \frac{zX(z) - z^{-1}X(z)}{2T_s}$$

$$\text{or } s \approx \frac{z - z^{-1}}{2T_s} \tag{2.3}$$

Considering a pole at $s = a$ gives the relation $z^2 - 2aT_s z - 1 = 0$. Therefore, (2.3) suggests that a pole at $s = a$ would map to a pole at $z = a \pm \sqrt{a^2 T_s^2 + 1}$. Therefore, an unstable pole ($a > 0$) in continuous-time system representation may result in a stable pole in discrete-time equivalent representation and vice versa using this approximation.

4. *Trapezoidal integration method*: Similar results can be obtained while approximating the integration operation. The integration operation is obtained by calculating area under the curve in a piece-wise manner. The area calculations based on the previous sample value or next sample value result in the same approximations as the difference equation approximations.

The trapezoidal integration method is illustrated in Figure 2.2. The piece-wise integration is calculated considering trapezium as a piece. The area under a trapezium is given by $T_s x[n-1] + \frac{T_s}{2}(x[n] - x[n-1])$ or $\frac{T_s}{2}(x[n] + x[n-1])$. The area of trapezium is added in the previous area calculation, and the discrete-time integration equation is given by $y[n] = y[n-1] + \frac{T_s}{2}(x[n] + x[n-1])$. Therefore, the transfer function of numerical integrator is

$$T(z) = \frac{T_s}{2} \frac{1 + z^{-1}}{1 - z^{-1}}$$

But, transfer function of the continuous-time integrator is $1/s$. Now, we have,

$$\frac{1}{s} \approx \frac{T_s}{2} \frac{1 + z^{-1}}{1 - z^{-1}}$$

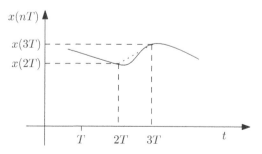

Figure 2.2 Illustration of trapezoidal integration

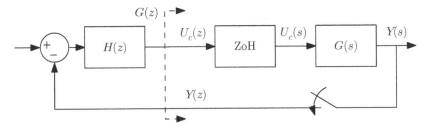

Figure 2.3 Exact discrete-time model of a continuous-time system

or

$$s \approx \frac{2}{T_s} \frac{1 - z^{-1}}{1 + z^{-1}} \tag{2.4}$$

Equation (2.4) suggests that a pole at a in s-plane would result in a pole at $(2 + aT_s)/(2 - aT_s)$. Therefore, an unstable pole ($a > 0$) in continuous domain would render $|2 + aT_s| > |2 - aT_s|$; therefore, it maps to an unstable pole in discrete domain using this approximation. For a stable pole ($a < 0$) in continuous domain, we get $|2 + aT_s| < |2 - aT_s|$. Therefore, it would always map to a stable pole in discrete domain. This approximation is a valid approximation for any value of sampling period. This approximation is also called as *Tustin Approximation*.

5. *Exact method*: The control system with an analog plant and embedded controller is shown in Figure 2.3. The analog plant is modeled in continuous-time and the transfer function is represented by $G(s)$. Similarly, the embedded controller is modeled in discrete-time and the transfer function is represented by $H(z)$. The output of the plant $y(t)$ is discretized and the corresponding discretized signal is $y[n]$. Figure 2.3 shows the corresponding Laplace transform $Y(s)$ and Z-transform $Y(z)$. Similarly, the discrete-time output of the controller $u_c[n]$ is applied to a ZoH (reconstruction filter) so that an equivalent analog signal $u_c(t)$ is applied to the plant. The Laplace transform of $u_c(t)$ is shown as $U_c(s)$, while the Z-transform of $u_c[n]$ is shown as $U_c(z)$. The objective is to find the transfer function $G(z)$ of the equivalent discrete-time system that includes sampling and reconstruction.

The plant is considered linear and shift-invariant. We can find the transfer function of the plant, if the response of plant to unit step is known. The discrete-time unit step signal $u_c[n]$ is chosen because the output of the reconstruction filter $u_c(t)$ is exactly a continuous-time unit step. Therefore, modeling process is as follows:

(a) Calculate step response of $G(s)$ ($Y(s) = G(s)/s$).
(b) Obtain output signal $y(t)$ by finding inverse Laplace transform of $Y(s)$.
(c) Find the sampled-time version of this step response, $y(nT_s)$.
(d) Find Z-transform $Y(z)$ of the output signal $y(t)$.

(e) Find $G(z)$ by dividing the step response $Y(z)$ by Z-transform of step $(z/(z-1))$.

The method described here incorporates the sampling and reconstruction processes.

Given the transfer function, the discrete-time controller design using Proportional-Integral-Derivative (PID) control is described next.

2.2 Discrete-time PID Controller Design

The controller design using PID technique is a very popular one. The reasons for its popularity are many; it is simple to implement, the literature is well established to support its stability claims and, more importantly, works in practice in majority of systems. Each system has nonlinear characteristics and is linearized at an operating point in a typical approach to design PID controller. The PID controller in these typical scenarios works for the linearized system in practice because the small changes due to nonlinear behavior act as disturbance and the PID controller is capable of rejecting these disturbances. The PID control concepts and their analysis have been developed exhaustively in literature. Since the implementation of an embedded controller is in discrete-time, the PID controller description is obtained from its continuous-time counterpart as follows.

Let the error between the reference and feedback signals be $e(t)$ and the output of controller be $u(t)$, the PID controller in the continuous-time form when initialized to $t = 0$ is described by

$$u(t) = K_p e(t) + K_i \int_0^t e(t)dt + K_d \frac{d}{dt} e(t) \qquad (2.5)$$

where K_p, K_i, and K_d are proportional, integral, and derivative gains. Here's a quick recap of actions possible with each of the terms, namely proportional control ($K_p e(t)$), integral control ($K_i \int_{-\infty}^{\infty} e(t)dt$), and derivative control ($K_d \frac{d}{dt} e(t)$). The proportional control is capable of following the changes in the reference signal (set-point), but any steady-state error needs to be rectified using the integral control. While integral control facilitates improving steady-state characteristics, the derivative control can improvise transient behavior.

Now, the PID controller description is needed in discrete-time. As a first effort, let us discretize (2.5) by computing the signals at sampled time instance. Now, the discrete-time controller output with sampling time T is given by

$$u(nT_s) = K_p e(nT_s) + K_i \sum_{k=0}^{n} e(kT_s) + \frac{K_d}{T_s}(e(nT_s) - e((n-1)T_s))$$

Or, in the form of difference equation, it is given by

$$u[n] = K_p e[n] + K_i \sum_{k=0}^{n} e[k] + \frac{K_d}{T_s}(e[n] - e[n-1]) \tag{2.6}$$

The term corresponding to the integration appears as accumulation ($\sum_{k=0}^{n} e[k]$). Defining an internal variable for accumulation helps in reducing the terms in (2.6). Let accumulation variable be $a[n]$ and given by $a[n] = \sum_{k=0}^{n} e[k]$. Now, we have $a[n-1] = \sum_{k=0}^{n-1} e[k]$ and therefore, (2.6) is now given by

$$u[n] = K_p e[n] + K_i a[n] + \frac{K_d}{T_s}(e[n] - e[n-1]), \text{where } a[n] = a[n-1] + e[n] \tag{2.7}$$

Now, the transfer function of discrete-time controller is obtained by taking Z-transform on both the sides of (2.7) and we get,

$$U(z) = K_p E(z) + K_i A(z) + \frac{K_d}{T}(E(z) - z^{-1}E(z)) \text{ with } A(z) = z^{-1}A(z) + E(z)$$

where $U(z)$, $E(z)$, and $A(z)$ are Z-transform of discrete-time signals $u[n]$, $e[n]$, and $a[n]$, respectively. Rearranging the terms, we get the controller transfer function

$$\frac{U[z]}{E[z]} = K_p + \frac{K_i}{1 - z^{-1}} + \frac{K_d}{T_s}(1 - z^{-1}) \tag{2.8}$$

Note that the controller transfer function given by (2.8) is obtained by backward difference method for differentiation, while the integral term uses accumulation method. The accumulation method when applied on differentiation renders results obtained from approximations using backward difference method. Let us look into the accumulation as applied on derivative term. Let the derivative control term in the PID controller output be given by $u_d(t)$ where

$$u_d(t) = K_d \frac{d}{dt}e(t)$$

Taking integration on both the sides gives

$$\int_0^{\infty} u_d(t)dt = K_d e(t)$$

The equivalent difference equation is now obtained as

$$\sum_{k=0}^{n} u_d[k] = K_d e[n] \tag{2.9}$$

Similarly, the relation at one previous sample is given by

$$\sum_{k=0}^{n-1} u_d[k] = K_d e[n-1]$$

Therefore, representing (2.9) in the form of one previous sample renders

$$u_d[n] + K_d e[n-1] = K_d e[n]$$

Or,

$$u_d[n] = K_d(e[n] - e[n-1])$$

which is nothing but the form obtained by approximation using backward difference method.

Remark 2.1   The controller design is in discrete-time as the controller is implemented in the embedded platform. Hence, the approximation methods are covered in Section 2.1. It is recommended to analyze the entire controller design in the discrete-time domain to avoid any implementation-related implications not captured at the time of designing the continuous-time controller.

Now, the section 2.3 describes the effects of sampling and quantization on performance of the embedded controller.

2.3   Stability in Embedded Implementation

When the signal is discretized, a significant amount of information is lost. But, is this information important? In particular, what is the price we pay by sampling? The answer to this question is analyzed by revisiting the concept of sampling.

2.3.1   Sampling

When the continuous-time signal $x(t)$ for $t \geq 0$ is sampled with sampling time T_s, the resulting discrete-time signal is represented by $x_n = x(nT_s) \ \forall \ n = 0, 1, 2, \dots$. The value of the signal at T_s time interval is recorded and the values between the sampling interval are discarded. The frequency spectrum of the sampled signal has the aliasing effect. If the continuous signals having frequencies $f \pm k/T_s \ \forall \ n = 0, 1, 2, \dots$ are sampled with sampling time T_s, the sampled signals are indistinguishable. Therefore, exact reconstruction is not possible unless the frequency of original signal is known a priori. The spectrum density of the discrete-time signal reflects the signal energy at the aliased frequencies as well. The Nyquist–Shannon sampling theorem suggests that the sampling frequency should be more than twice the bandwidth of the signal to avoid the overlapping of energies due to aliasing.

In control systems, the bandwidth of the system is typically low and sampling rate criteria is achieved. Therefore, frequency aliasing is not a problem in embedded controllers. Reconstructing the signal is anyway not an aim in embedded controllers. The concern is the injection of sensor noise before sampling. Feedback signal would contain high-frequency noise components that may be aliased to the low-frequency zone and can be within the bandwidth of the control system. The aliasing of sensor noise in the system's bandwidth can be avoided by incorporating *anti-aliasing filter* or a low-pass filter before sampling. We know that a filter with significant roll-off produces high-phase shifts at low frequencies. The output of a mobile robotic system is typically a low-frequency signal. Therefore, the introduction of a low-pass filter would introduce a significant phase shift in the signal, which would reduce the stability margin of control system. Let us understand this interesting and important aspect of sampling from the perspectives of embedded controller design in detail.

The ZoH, used for reconstructing the signal, holds the value of most recent sample and holds it for the sample time. It also acts as a low-pass filter with the frequency amplitude response given by

$$A = \sin c(\pi T_s f) = \begin{cases} 1 & f = 0 \\ \frac{\sin(\pi T_s f)}{\pi T_s f} & \text{otherwise} \end{cases}$$

A typical response of the ZoH circuit is shown in Figure 2.4. The response shows that the high-frequency components would be attenuated before reaching the system or plant. In video and image processing, this attenuation would result in blurring, but rejection of high-frequency component is desirable in control system.

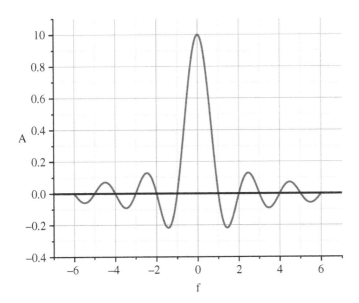

Figure 2.4 Nature of frequency amplitude response of ZoH

A simple ZoH also generates significant noise. A low-pass filter with high cut-off frequency would eliminate this noise and smoothen the DAC (Digital-to-Analog Converter) output. However, an added filtering results in addition of phase shift. Therefore, the stability of the control system must be analyzed after incorporating the phase shift introduced by the low-pass filter to avoid effects cropped up due to sampling.

Another integrated problem with sampling is the sampling jitter. No clock generation is accurate. The deviation in sampling instances due to imprecise sampling clock is termed as sampling jitter. The embedded technology is now advanced and produces minor jitters that can be ignored for mobile robotic applications. But it is still important to consider the jitter when multiple sub-systems of a robotic system use different embedded platforms and hence different sources of clock generation. Let the inaccuracy in the sampling at nth discrete instance be ϵ_n. The signal value at nth discrete instance is given by

$$x[n] = x(nT_s + \epsilon_n)$$

$$= x(nT_s) + \epsilon_n \frac{d}{dt} x(t) \Big|_{t=nT_s} + \epsilon_n^2 \frac{d^2}{dt^2} x(t) \Big|_{t=nT_s} + \cdots$$

$$\approx x(nT_s) + \mathcal{N}_n$$

where $\mathcal{N}_n = \epsilon_n \frac{d}{dt} x(t) \Big|_{t=nT_s}$ is called sampling jitter. Note that the sampling jitter not only depends on the imprecision in sampling clock, it also depends on the time derivative of the signal at the sampling instant. A fast system with controller having high gain to instantaneous changes (high derivative gain) would generate a significant noise due to sampling jitter. But a slow system with a low proportional gain and high integral gain controller may not couple sampling jitter.

2.3.2 Quantization

The embedded controller is interfaced with the digitized (or quantized discrete) signal. When the signal is represented in binary form and the values are quantized to the nearest quantization level, the quantization results in *round-off* error, whereas if the signal values are quantized to the lower quantized level, the quantization results in *truncation* error. Therefore, quantization error is either *round-off* or *truncating*. The quantization error does not propagate to the next sample value in the closed-loop control system. Therefore, it is difficult to analyze the effect of quantization error. However, it is possible to analyze if it is considered as a noise generated through the controller. Following three approaches are used to analyze the effect of quantization:

1. Obtain upper bound on the errors that can be caused by quantized effects.
2. Neglect transients and analyze only the steady-state effects.

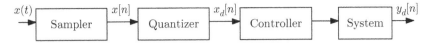

Figure 2.5 An open-loop control system with quantizer

3. Treat the quantization error as noise arising from stochastic sources and proceed with a statistical analysis.

Next, each approach is described in detail.

1. *Worst-case analysis to calculate upper bound*: The open-loop embedded control system is shown in Figure 2.5. The output signal of the sampler is represented by $x[n]$. This signal $x[n]$ is the discrete-time signal of the continuous-time signal $x(t)$. The digitized signal $x_d[n]$ obtained after the quantization is given by $x_d[n] = x[n] + q[n]$, where $q[n]$ is the quantization error at nth discrete instance. The system output $y[n]$ is the response of the plant to the digitized signal $x_d[n]$. If the Z-transform of $x_d[n]$ is $X_d(z)$ and $y[n]$ is $Y(z)$, then the relationship $Y(z) = G(z)H(z)X_d(z)$ holds, where $G(z)$ is the discrete transfer function of the plant and $H(z)$ is the discrete transfer function of the controller. Next, we obtain a discrete transfer function of the open-loop control system as shown in Figure 2.5 using the method described in Section 2.1. Let the response of the plant for a discrete-time signal $x[n]$ be $y[n]$, then we have

$$y[n] = y_d[n] + y_q[n] \tag{2.10}$$

where $y_q[n]$ is the error in output due to quantization error. Note that (2.10) holds because the system under consideration is linear and shift-invariant. The system is linear; therefore, we have

$$Y_q(z) = G(z)H(z)Q(z)$$

where $Y_q(z)$ and $Q(z)$ are the Z-transform of signals $y_q[n]$ and $q[n]$, respectively. Let the open-loop transfer function $G(z)H(z)$ be $T_{op}(z)$. Now, the output sequence due to quantization error is given by

$$y_q[n] = \sum_{m=0}^{n} t_{op}[m]q[n-m] \tag{2.11}$$

where $t_{op}[n]$ is the inverse Z-transform of $T_{op}(z)$ or the impulse response of open-loop system. If the weight of the least significant bit is δ_q, then the absolute value of quantization error is bounded by $\delta_q/2$ for rounding and δ_q for truncation. Therefore, equation (2.11) gives the upper bound of output error due to quantization as

$$|y_q[n]| \le \sum_{m=0}^{n} |t_{op}(m)||e(n-m)|$$

or

$$|y_q[n]| \leq \sum_{m=0}^{n} |t_{op}(m)|\alpha \tag{2.12}$$

where

$$\alpha = \begin{cases} \delta_q/2 & \text{for truncation} \\ \delta_q & \text{for rounding} \end{cases}$$

Therefore, the upper bound of the output error in open-loop system depends on the system and controller models.

The analysis on closed-loop system is also similar. The closed-loop system with quantization error source considered as an additive source is shown in Figure 2.6. The quantization is considered in the feedback loop for this system. The input signal is also quantized, but the effect of the quantization error at the input can be analyzed in the similar manner.

Since the control system is a linear and shift-invariant system, the output signal contributed by the quantization error is obtained by considering zero input ($x_d[n] = 0$). This results in the transfer function

$$T_{cl}(z) = \frac{Y_d(z)}{Q(z)} = \frac{-G(z)H(z)}{1 + G(z)H(z)}$$

The upper bound on the output due to quantization error is given by

$$|y_d[n]| \leq \sum_{m=0}^{n} |t_{cl}[m]||q[n-m]|$$

where $t_{cl}[n]$ is the inverse Z-transform of $T_{cl}(z)$. Next, the steady-state behavior of the output due to quantization error is analyzed.

2. *Steady-state analysis*: We know that the maximum value of quantization error $q[n]$ is α and the upper bound on the output due to quantization error is given by (2.12). The final value theorem suggests that the steady-state value of the output $y_d[n]$ is bounded by

$$y_d[n]|_{n\to\infty} = \lim_{z\to 1}(z-1)T(z)\frac{\alpha}{1-z^{-1}}$$

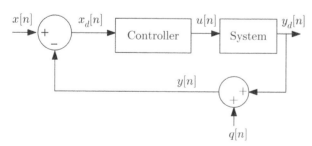

Figure 2.6 A closed-loop system with quantizer

where $T(z) = T_{op}(z)$ for open-loop control system and $T(z) = T_{cl}(z)$ for closed-loop control system. Therefore, the steady-state output due to quantization error is bounded by

$$y_d[n]|_{n \to \infty} \leq \alpha T(z)|_{z \to 1}$$

3. *Stochastic analysis*: The deterministic analysis can give the upper bound on the output due to quantization error. Further analysis is not possible, as the error characteristic is unknown. Therefore, stochastic approach is used to further analyze the output characteristic due to quantization. The quantization error is considered as a random variable and the quantization process is considered as a noise generation process. This assumption is valid because the quantization error is bounded by $(-\delta_q/2, \delta_q/2)$ for rounding and $(0, \delta_q)$ for truncation, and the occurrence of an error value between this interval is equally likely.
The average value, μ of the quantization noise is given by

$$\mu = \mathrm{E}\langle q \rangle = \begin{cases} \delta_q/2 & \text{for truncation} \\ 0 & \text{for rounding} \end{cases} \tag{2.13}$$

The variance of noise for rounding is given by

$$\mathrm{E}\langle q^2 \rangle = \frac{1}{\delta_q} \int_{-\delta_q/2}^{+\delta_q/2} q^2 dq$$

$$= \frac{1}{\delta_q} \frac{q^3}{3} \Big|_{-\delta_q/2}^{+\delta_q/2}$$

$$= \frac{\delta_q^2}{12} \tag{2.14}$$

Similarly variance of noise for truncation is given by

$$\mathrm{E}\langle q^2 \rangle - (\mathrm{E}\langle q \rangle)^2 = \frac{1}{\delta_q} \int_0^{\delta_q} q^2 dq - \left(\frac{\delta_q}{2}\right)^2$$

$$= \frac{1}{\delta_q} \frac{q^3}{3} \Big|_0^{\delta_q} - \frac{\delta_q^2}{4}$$

$$= \frac{\delta_q^2}{3} - \frac{\delta_q^2}{4} = \frac{\delta_q^2}{12} \tag{2.15}$$

Therefore, the variance, σ^2 of noise is given by

$$\sigma^2 = \frac{\delta_q^2}{12} \tag{2.16}$$

for both truncation and rounding.
The Z-transform of output due to quantization error is given by

$$Y_d(z) = T(z)Q(z) \tag{2.17}$$

where $T(z) = T_{op}(z)$ for open-loop control system and $T(z) = T_{cl}(z)$ for closed-loop control system. The mean value of the signal $y_d[n]$ is obtained as follows: At the steady-state, $y_d[n + 1] \approx y_d[n]$; therefore, average value at the steady-state is given by

$$E\langle y_d \rangle = T(z)|_{z \to 1} E\langle q \rangle$$

The variance of output due to quantization noise is calculated as follows: The inverse Z-transform of equation (2.17) gives

$$y_d[n] = \sum_{m=0}^{n} t[m]q[n - m] \tag{2.18}$$

The second central moment for the output signal is given by

$$
\begin{aligned}
E\langle (y_d[n])^2 \rangle &= E\left(\sum_{m=0}^{n} t[m]q[n - m] \sum_{k=0}^{n} t[k]q[n - k] \right) \\
&= E\left(\sum_{m=0}^{n} \sum_{k=0}^{n} t[m]t[k]q[n - m]q[n - k] \right) \\
&= \sum_{m=0}^{n} \sum_{k=0}^{m} t[m]t[n]E\left(q[n - m]q[n - k] \right)
\end{aligned}
$$

We know that

$$E\left(q[n - m]q(n - k) \right) = (\sigma^2 + \mu^2)\delta[m - k] \tag{2.19}$$

where σ^2 is the variance of quantization noise defined by (2.16), μ is the mean value of quantization noise defined by (2.13), and $\delta[n]$ is the impulse at $n = 0$. From (2.18) and (2.19), we get

$$E\langle (y_d[n])^2 \rangle = (\sigma^2 + \mu^2) \sum_{m=0}^{n} (t[m])^2$$

This shows that the quantization noise has the same effect as the gain of the system.

The Section 2.3.3 describes the effect of processing time on the performance of the control system.

2.3.3 Processing Time

When the embedded controller implements the functionality, a considerable amount of time may be required for execution. The total time required for processing sensor information, executing functionality, and actuating the plant (includes response time of the system) decides the loop time of the control system. For the embedded controller implementation, it is required to ensure that sampling time must be greater than the loop time of the single rate control system. In other words, the time required for processing sensor data and subsequently generating the controller output must be less than the response time of the system (or actuators). With the advancement of embedded technology, this requirement can be easily attained and does not introduce any significant delay in the system.

In the worst-case, any significant processing time due to sophisticated sensor processing or controller functionality is modeled as delay for controller design or stability analysis.

2.4 Notes and Further Readings

This chapter gives a systematic approach for discrete-time controller formulation for a continuous-time system (a mobile robot). Once the representation in discrete-time is obtained, the controller design using PID technique is discussed in detail. The objective of this book is to address the embedded implementation of discrete-time controller; therefore, this chapter covers the effect of digitization too. The effect of concerned issues like sampling, quantization, and processing time on the controller performance is discussed. It is advised to cover stability of discrete-time controller for designing and tuning the PID gains from any standard material. A few recommendations for the text-book material on discrete-time control system can be seen in Ogata (1987) and Rabbath and Léchevin (2014).

Next, the concepts of discrete-time control design for embedded implementation are applied on a system for better understanding of these concepts. The system description is motivated by the mobile robotic applications in this book. The mobile robotic applications require a few common tasks and computations to be performed repetitively. These tasks and computations with their embedded implementation are discussed in Chapter 3. The embedded solutions for performing complex computations are also presented.

3

Embedded Control and Robotics

Various robotics applications are being explored in many domains such as defense, space, manufacturing, agriculture, and many more. Be it any robot; the high-level task planning applies the concept of partitioning the problem into multiple sub-tasks. These sub-tasks may or may not be independent. For example, collision detection and avoidance is an essential sub-task in most of the robotic applications. It may be independently executing until a collision possibility is detected. Each of this sub-task is in itself a control system interfaced with other systems. Figure 3.1 describes a typical methodology of decomposing into sub-tasks to achieve the robotic application's objectives.

As shown in Figure 3.1, a high-level planner generates commands to move the robot to achieve the robotic application's objectives. These commands are in the form of a navigation query. The navigation query describes the robot movements from one state to another. The state of the robot is its position, orientation, or both along with their derivatives. Once the robot's final state is available through the navigation query, the next step is to plan its path. The path planning targets avoiding collision with obstacles. These obstacles may be known a priori, or their information is computed on-the-go while sensing the environment using sensors mounted on the robot.

After planning the high-level task and further path or trajectory, appropriate commands are generated for the robot by applying transformations specific to the robot. The localization technique then captures the robot's movement. The outcome of the localization technique is the robot's coordinates, its orientations, or both with respect to the environment. The robot's state as feedback refines the high-level plan and path-planning stages. As described by the localization, the state of the robot acts as a feedback to the planning stages. It is worth noting that more or less each sub-task shown in Figure 3.1 is a control system in itself. For example, the localization block takes feedback from its output as a present state and input as the robot's control commands or encoder information along with other sensor data to calculate the next state.

Embedded Control for Mobile Robotic Applications, First Edition.
Leena Vachhani, Pranjal Vyas, and Arunkumar G. K.
© 2022 The Institute of Electrical and Electronics Engineers, Inc. Published 2022 by John Wiley & Sons, Inc.
Companion website: www.wiley.com/go/vachhani/embeddedcontrolforroboticapp

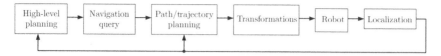

Figure 3.1 Autonomous robot planning as a control system

This chapter further presents a detailed explanation of these sub-tasks, formulating the embedded implementation requirements for typical methods corresponding to each sub-task. This chapter's approach is to extract the requirement of a method such that it is suitable for implementation in an embedded platform. Further, an approach to design a multi-agent scenario, where multiple robots collaborate to complete a task effectively, is discussed. This approach caters to the scenarios where each agent is small in size and controlled by an embedded platform with limited resources.

3.1 Transformations

The robot's body rotates and translates with respect to a reference point (refer point *M* in Figure 1.4). The processing of the sensor information to obtain the robot's states requires computations with respect to a reference point on the robot. For example, a depth sensor is mounted at a location different from the reference point on the robot. Now, the depth sensor provides distance to the obstacle in its field of view with respect to the point where it is mounted. But, high-level planner needs computations with respect to a reference point. These computations typically need transformation in terms of 2D and 3D translation and rotation. The discussed scenario of transformation of sensing information to the reference point requires fixed rotation and translation to compute; there are many other sub-tasks that need transformations as elementary computations. One of these sub-tasks is accounting for the finite size of the robot in collision avoidance and high-level planning.

These 2D and 3D transformations are introduced in this chapter, and various ways of computing it in embedded platform are discussed in Chapter 4.

3.1.1 2D Transformations

A point $P(x, y)$ in 2D space in vector form is represented as $[x \quad y \quad 1]^T$

Translation of a point $P(x, y)$ in a 2D plane is carried out by adding the displacement along x-axis and y-axis, respectively. The new coordinates of point P are given by

$$x' = x + x_d \tag{3.1}$$

$$y' = y + y_d \tag{3.2}$$

where x' and y' are the new coordinates and x_d and y_d are displacement along x and y axes, respectively. Therefore, the transformation for translation is given by

$$\begin{bmatrix} x' \\ y' \\ 1 \end{bmatrix} = \begin{bmatrix} 1 & 0 & x_d \\ 0 & 1 & y_d \\ 0 & 0 & 1 \end{bmatrix} \begin{bmatrix} x \\ y \\ 1 \end{bmatrix}$$

where the corresponding transformation is given by

$$T = \begin{bmatrix} 1 & 0 & x_d \\ 0 & 1 & y_d \\ 0 & 0 & 1 \end{bmatrix} \begin{bmatrix} x \\ y \\ 1 \end{bmatrix}$$

Let us consider a scenario where sensor reference point is $P(x_p, y_p)$ on the robot's body with respect to the body reference for robot at point M (origin). The corresponding positions at nth sampling time instance are shown in Figure 3.2a. The distance between points P and M is S, and the angular displacement of point P with respect to the forward direction of the robot is θ. In other words, the polar

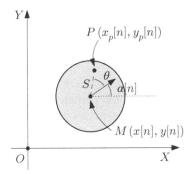

(a) Sensor reference point P with respect to robot's body frame of reference

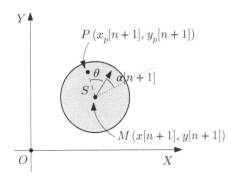

(b) Rotation of robot's body and sensor reference point

Figure 3.2 Rotation in 2D

coordinates of point P with respect to body frame of reference at point M are (S, θ) that are fixed.

Let the robot turn by an angle $\phi[n]$ at nth time instance at its own place and the new location of the robot is given by $(x[n+1], y[n+1])$ that is same as previous location $(x[n], y[n])$, but the orientation is changed to $\alpha[n+1] = \alpha[n] + \phi[n]$ as shown in Figure 3.2b.

Now, the coordinates of sensor reference point P can be obtained in two different ways.

1. *Using new coordinates of robot's reference M:*

$$\begin{bmatrix} x_p[n+1] \\ y_p[n+1] \end{bmatrix} = \begin{bmatrix} x[n+1] + S\cos(\alpha[n+1] + \theta) \\ y[n+1] + S\cos(\alpha[n+1] + \theta) \end{bmatrix} \tag{3.3}$$

2. *Using previous coordinates of sensor reference $P\left(x_p[n], y_p[n]\right)$:*

$$\begin{bmatrix} x_p[n+1] \\ y_p[n+1] \end{bmatrix} = R_{\phi[n]} \begin{bmatrix} x_p[n] \\ y_p[n] \end{bmatrix} \tag{3.4}$$

where

$$R_{\phi[n]} = \begin{bmatrix} \cos(\phi[n]) & \sin(\phi[n]) \\ -\sin(\phi[n]) & \cos(\phi[n]) \end{bmatrix} \tag{3.5}$$

Many a time sensor reference computation is preferred with respect to its own previous coordinates because environment information is obtained from the sensor. In other words, the environment information is coded with respect to the sensor reference point hence, the information update is easier once it is related to its own reference at the previous instance.

Transformation of robot in 2D involves both translation and rotation. Rotation of the reference point on the robot; involves translation of origin to the reference point using transformation T by $(x_d, y_d) = (x[n+1] - x[n], y[n+1] - y[n])$, followed by rotation using transformation $R_{\phi[n]}$ (recall $\phi[n] = \alpha[n+1] - \alpha[n]$). These transformations in 3D are now discussed in Section 3.1.2.

3.1.2 3D Transformations

Coordinate rotation in 3D is helpful in many applications which involve motion in 3D space specially in aerial and underwater robots. Like the case in 2D where the sensor reference point undergoes 2D transformations, the 3D transformations are used when the world is represented in 3D. Let us formulate the problem of translation and rotation in 3D for the case when the point P is on an object and reference of point P is with respect to the moving robot. The rotation and translation of reference frame for a stationary point P is same as rotation and translation of point P with respect to stationary reference frame but in the opposite direction.

A point $P(x, y, z)$ in 3D in its vector form is given by $[x \quad y \quad z \quad 1]^T$. Similar to the 2D case, translation of point $P(x, y, z)$ by a displacement of x_d, y_d, z_d is obtained by the transformation $T_{x_d y_d z_d}$ using the following operation

$$
\begin{bmatrix} x' \\ y' \\ z' \\ 1 \end{bmatrix} = \begin{bmatrix} 1 & 0 & 0 & x_d \\ 0 & 1 & 0 & y_d \\ 0 & 0 & 1 & z_d \\ 0 & 0 & 0 & 1 \end{bmatrix} \begin{bmatrix} x \\ y \\ z \\ 1 \end{bmatrix} = T_{x_d y_d z_d} \begin{bmatrix} x \\ y \\ z \\ 1 \end{bmatrix}
$$

As explained in Section 1.2.2, rotations with respect to X–Y–Z axis depend on the sequence in which it is rotated with respect to each axis. For consistency in notation, we redefine the transformation corresponding to the rotation about Z, Y, and X axes by angles α, β, and γ are as follows:

$$
R_{z,\alpha} = \begin{bmatrix} \cos\alpha & \sin\alpha & 0 & 0 \\ -\sin\alpha & \cos\alpha & 0 & 0 \\ 0 & 0 & 1 & 0 \\ 0 & 0 & 0 & 1 \end{bmatrix}
$$

$$
R_{y,\beta} = \begin{bmatrix} -\sin\beta & \cos\beta & 0 & 0 \\ 0 & 0 & 1 & 0 \\ \cos\beta & \sin\beta & 0 & 0 \\ 0 & 0 & 0 & 1 \end{bmatrix}
$$

$$
R_{x,\gamma} = \begin{bmatrix} 0 & 0 & 1 & 0 \\ \cos\gamma & \sin\gamma & 0 & 0 \\ -\sin\gamma & \cos\gamma & 0 & 0 \\ 0 & 0 & 0 & 1 \end{bmatrix}
$$

Rotation of a point $P(x, y, z)$ along an arbitrary axis involves a combination of these translation and rotation matrices. The steps of rotation about an arbitrary axis L by an angle α are as follows:

1. Translate the origin to point A. Point A is the orthogonal projection of point P on the axis L.
2. Rotate vector L with respect to Y axis so that it lies in yz-plane. The angle of rotation is β.
3. Rotate vector L with respect to X axis so that the vector aligns along Z axis. The angle of rotation is γ.
4. Now the vector gets aligned to Z-axis. The rotation of the point P can be performed with respect to Z-axis by angle α.
5. After performing the desired rotation, transformation is performed in the reverse order.

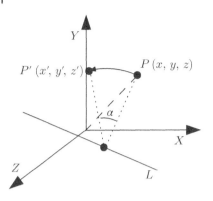

Figure 3.3   Rotation in 3D

The overall transformation involves the following matrix operations: $R_{L,\alpha} = T_{-A_x,-A_y} \times R_{y,\beta} \times R_{x,\gamma} \times R_{z,\alpha} \times R_{x,-\gamma} \times R_{y,-\beta} \times T_{A_x,A_y}$, where the (A_x, A_y) are the coordinates of point A with respect to origin and α, β, and γ are the rotation about X, Y, and Z axes as illustrated in Figure 3.3.

The transformations can be implemented as matrix multiplication; however, there is another interesting and efficient method of implementing vector rotations in embedded platform. This alternate method uses concepts of CO-ordinate Rotation DIgital Computer (CORDIC) explained in Chapter 4.

3.2   Collision Detection and Avoidance

Local sensing plays an important role in changing the trajectory of a mobile robot after encountering an obstacle. The sensors such as proximity, vision, and LiDaR (Light Detection and Ranging) are typically used in mobile robotics to perceive the local environment. Proximity sensors such as IR, ultrasonic sensors, and LiDaR detect the distance and sometimes direction to the nearest obstacles. Vision sensors are used for obstacle detection and distance to the obstacle using techniques such as stereo vision and depth estimation. The robot is said to be on the course of collision with other obstacle if the samples within a time horizon detect the collision. Hence, in many algorithms, it is needed to estimate velocity as well. The collision detection and avoidance techniques are primarily developed based on the sensing capabilities of the mobile robot. For example, using the proximity or vision sensors is one of the cost-effective ways for detecting the objects as compared to that of LiDaR. However, for determining the exact position of the object using proximity or vision sensors, more than two sensors are deployed. We present some of the popular collision avoidance techniques regardless of the specific sensor capability. These techniques are described from the perspective of embedded processing and computing requirements for collision detection and avoidance.

3.2.1 Vector Field Histogram (VFH)

The Vector Field Histogram (VFH) technique of Johann and Koren (1991) and Kazem et al. (2010) creates a local map for the robot with the help of local sensors. This is in the form of occupancy grid which is represented in the form of polar histogram. The X-axis represents the angles covering 360°, and Y-axis represents the probability of having an obstacle in that direction. The regions with low probability values are identified which represent the open gaps where the robot can steer to avoid the obstacles. Let the direction corresponding to the open gap be given by θ_{open} with reference to the robot's reference frame as shown in Figure 3.4. The robot's reference frame is located at point M with orientation $\alpha[n]$ (forward direction of the robot at nth instance) with respect to the world frame. Now, the angle difference between the goal direction and forward direction is θ_{goal}.

Minimizing a cost function based on goal direction θ_{goal}, the representative direction of open gap θ_{open} and steering amount to move in new direction $(\alpha[n+1] - \alpha[n])$ is used for determining the direction in next time instance. The cost function C is given by

$$C = w_1 \cdot \theta_{goal} + w_3 \cdot (\theta_{open}) + w_2 \cdot (\alpha[n+1] - \alpha[n]) \tag{3.6}$$

where w_1, w_2, w_3 are respective weights for various cost components.

The weights of the cost function C are used to tune the behavior of the robot. The VFH+ algorithm is the improvement over the VFH technique which takes into account the kinematic constraints of the robot while computing the possible gaps. For embedded implementation of VFH technique, following modular functions are suggested:

1. Maintain a local map of probability distribution using sensor data and find the open gaps.
2. Develop an iterative algorithm to find cost function value for incremental values of discrete angles.
3. Take decision based on the computed cost function.

Figure 3.4 Vector field histogram

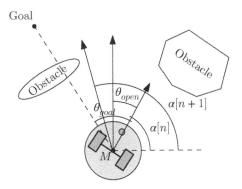

3.2.2 Curvature Velocity Technique (CVM)

The Curvature Velocity Technique (CVM) technique of Simmons (1996) and Yong and Simmons (1998) assumes that the robot travels only along the circular arcs. This technique takes into account the kinematic and dynamic constraints for the robot movement. The obstacles are assumed to be circular in shape. The physical constraints related to the robot's movement and obstacle are then mapped into velocity space. Based on the bounds linear speed $|v|$ and angular speed $|\omega|$, expressed as $v_{min} < |v| < v_{max}$ and $\omega_{min} < |\omega| < \omega_{max}$, the maximum and minimum curvatures of the robot are obtained. Note that curvature is obtained using $c = (\omega/v)$. However, the curvatures bounds c_{min} and c_{max} that collide with an obstacle are computed as depicted in Figure 3.5. The control inputs that are linear and angular velocities corresponding to non-colliding curvatures are computed by considering both the kinematics and dynamic of the robot so that the robot can operate at high speeds avoiding the collision. The method attempts to maximize the forward travel of the robot in comparison to other techniques.

3.2.3 Dynamic Window Approach (DWA)

The Dynamic Window Approaches (DWAs) of Fox et al. (1997), Ogren and Leonard (2005), and Seder and Petrovic (2007) are another set of techniques which take into account the kinematic constraints of the robot. The dynamic window works on creating a velocity space and thus mapping the obstacle's equivalent in the velocity space. The velocity space is the set of all the possible linear and angular speeds possible for the robot. This technique first computes a *dynamic window* which is the set of all the possible (v, ω) pairs that can be achieved by the robot in the next time instant. This dynamic window is then reduced to keep only those (v, ω) pairs which result in no collision for the next time instant. The new direction of motion is computed by optimizing an objective function over all admissible (v, ω) pairs. The objective function is calculated based

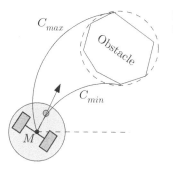

Figure 3.5 Illustration of curvature velocity technique

on heading toward the goal, forward velocity of the robot, and distance from the closest obstacle given by

$$C = w_1 \cdot heading(v, \omega) + w_2 \cdot distance(v, \omega) + w_3 \cdot velocity(v, \omega) \qquad (3.7)$$

where w_1, w_2, and w_3 are weights corresponding to the heading, distance, and velocity functions. The $heading(v, \omega)$ is alignment of the robot with the target direction. For this, the target direction is computed based on the predicted position. The predicted position is computed by selecting a particular velocity in the next time interval. The function $distance(v, \omega)$ represents the distance to the closest obstacle which intersects with the curvature. The function $velocity(v, \omega)$ computes the progress of the robot on the corresponding trajectory. The combination of all these three components allow the robot to avoid the collision while quickly progressing toward the goal by following a smooth trajectory. However, this approach focuses on taking local decisions for the dynamic window for the current instant.

3.3 Localization

Localization is the most important building block of mobile robot navigation where the robot has to determine its position and/or orientation (state of the mobile robot) with respect to the environment. A sensor that can directly give the position information is global positioning system (GPS). However, the commercial-grade GPS provides accuracy about the position upto only several meters, which is insufficient and unacceptable for various indoor and outdoor applications. Furthermore, localization of a robot is not only about knowing absolute position but also about its relative position with respect to target humans and obstacles. The main challenges faced in localization are noise and aliasing from sensing.

Sensors are an integral part of localization. They perceive the environment and create an environment representation or map. However, the noise induced due to environmental aberrations may result in wrong interpretation of the robot's location. Any wrong interpretation on the robot's location would result in issuing wrong commands by the controller to the robot and leads to the failure. For instance, if the map of an indoor environment is created from vision sensor, the robot may use its unique color values for localizing itself. If the illumination of the environment changes (say due to weather change or any environmental factor), then ensuring uniqueness of color identification may not be possible due to noisy image pixel values. Similarly, there can be erroneous readings when ultrasonic sensors are used for gathering environment information and, subsequently, in estimating the robot's location. When the ultrasonic sensor emits sound wave, the amount of reflected/received signal energy depends on the material characteristics and surface reflection. Thus, reflections from various materials would result in erroneous sensor readings. Moreover, mobile robots typically contain array of multiple sonar sensors. The array of multiple sensors cover a Field-of-View or scan region in the robot's neighborhood. Use of multiple sensors occasionally causes multipath interference between sonar emissions resulting in large errors.

Sensor aliasing is another major problem which can occur while localizing because of non-uniqueness of sensor readings. For example, an ultrasonic sensor provides only range information without any additional information about the type of material, color, texture from which the sonar echo is reflected. Mobile robots use multiple sonar sensors whose similar readings can result in different states in the map, later, creating a confusion of being in more than one position for the robot. Even with noise-free sensors, single reading from them is still insufficient to correctly predict the robot's location.

Challenges in localization not only lie in sensor noises and aliasing effects, but also because of noises present in actuators such as wheels of the mobile robot and end effector of a manipulator. In mobile robot, odometry (wheel encoders), and dead reckoning (heading sensors) are frequently used for tracing the movement of the robot. The information from wheel encoders and heading sensors are integrated with time to get the position information. The error gets accumulated over time due to various factors such as wheel misalignment, unequal wheel diameter, variation in wheel contacts, and limited resolution of timers and measurement. Therefore, other localization methods must be used to update the position from time to time, and it is desirable to develop an error model for odometry accuracy.

Belief representation is a concept in which a robot develops a belief regarding its location with respect to the map of the environment. Typically there are two techniques used for belief updates of the robot's position: single-hypothesis belief and multiple-hypothesis belief. In single-hypothesis belief, the robot's location is considered to be as a single unique point. If the robot's motion results in large uncertainty due to actuator or sensor noises, then single-hypothesis belief could also render erroneous results. In multiple-hypothesis belief, the robot considers a set of locations for its probable location. These sets of locations are based on an arbitrary probability distribution model. The most popular probability distribution used for estimation is Gaussian distribution with mean μ and standard deviation σ. Thus, in multiple-hypothesis belief robot explicitly carries uncertainty about its own location. By acquiring information from sensors, the robot corrects its belief and subsequently decreases the uncertainty in the robot's correct location. The challenge involved with multiple-hypothesis belief is that the robot has to perform decision making based on hypothesis on multiple locations. Decision making on multiple-hypothesis could result in involved computations.

Many popular localization techniques are used in mobile robotics such as Markov localization Fox et al. (1999) and Fox et al. (1999b), Kalman Filter localization Jetto et al. (1999) and Kwon et al. (2006), Monte Carlo Localization Dellaert et al. (1999), Thrun et al. (2001), Fox et al. (1999a), and Kümmerle et al. (2008), Unscented Kalman Filter Localization Ullah et al. (2019), Sun et al. (2010), and Lin et al. (2020). We discuss a general solution strategy for global localization using Markov technique to form a basis for understanding probabilistic techniques.

Consider a mobile robot in a known environment. It is possible to track the robot's movement using either odometry or a motion model. This type of direct measurement of the location induces uncertainty over time. This uncertainty in location keeps on growing as the robot moves. To prevent the uncertainty from growing unbounded, the robot controller uses information collected from on-board sensors (ultrasonic sensors, vision sensors, and LiDaR sensors, etc.) to make observations and predict the robot's location from the map of the known environment. The location predicted from odometry is combined with the sensor information to get a better estimate on the robot's location. The location estimation mainly consists of two steps: first, *prediction* of the robot's location is obtained using odometry or a robot's motion model based on control inputs, and second, an *update* step is used to correct the location estimation during the prediction step using the sensor information. Since the estimated locations using this localization technique carry uncertainty, the use of probability operations is inherent.

Let $X[n] = (x[n], y[n], \alpha[n])$ denote the robot's position and orientation (location in the environment/state of the robot) for the planar motion and $u[n]$ be the control input vector at sample time instance n. If the initial robot position p_0 is known, then the robot path can be easily integrated based on previous locations and control inputs in the absence of noise. Sensors such as camera, range, and LiDaR sensors mounted on the robot perceive the environment. The outputs returned by the sensor referred to as measurements/observations are denoted as $z[n]$ at n. Finally, let the map of the environment be denoted as vector M.

The robot's state/location is estimated based on estimated *beliefs*. The beliefs are propagated over time using the *prior* belief (prediction step) and the *posterior* belief (update/correction step). When the current robot position is predicted based on the present and past control inputs (i.e. $u[1], u[2], \ldots, u[n]$ denoted by $u[0 : n]$) and past observations ($z[1], z[2], \ldots, z[n-1]$ denoted by $z[0 : (n-1)]$), then the *prior* belief $\overline{bel}(X[n])$ is given by

$$\overline{bel}(X[n]) = p(X[n] \mid z[1 : (n-1)], u[1 : n])$$

The *posterior* belief $bel(X[n])$ computed after the current observation from sensors $z[n]$ is taken into account along with the past observations and current and past control inputs and is given by

$$bel(X[n]) = p(X[n] \mid z[1:n], u[1:n])$$

The computation of posterior belief is called correction/updation step, since the robot pose is corrected after observation.

Among all the localization techniques, Markov localization is the most common. Markov localization assumes that the current state $X[n]$ depends just on the previous state $X[n-1]$ and the current control input $u[n]$ and measurement $z[n]$. Therefore based on this assumption, the *prior* belief computed for the current robot position is dependent on the previous belief $bel(X[n-1])$ and its odometric input. Hence,

$$\overline{bel}(X[n]) = \sum p(X[n] \mid u[n], X[n-1]) \; bel(X[n-1])$$

It follows from the theorem of total probability, which integrates over all the possible ways in which robot position $X[n]$ could be reached. Similarly, the *posterior* belief corrects the predicted position based on the current measurements $z[n]$ and former belief state $\overline{bel}(X[n])$ using *Bayes rule*:

$$bel(X[n]) = \frac{p(z[n] \mid X[n], M) \; \overline{bel}(X[n])}{p(z[n])}$$

For an efficient embedded implementation, a few tricks are applied to reduce the computations for updating belief distribution over all the states/locations in the environment. For each hypothesis, the belief updation is incorporated with beliefs of nearby locations only. This results in drastic reduction in computations as well as truncation of very low beliefs to zero becomes straightforward. An extra step of normalization further adjusts the range of values that a limited number of data-bits can accommodate.

In the Section 3.4, we briefly describe the popular path- planning techniques used for navigation.

3.4 Path Planning

In mobile robotics, path planning is well-studied problem and many algorithms have been developed for path planning. The concept of configuration space developed originally from the study of industrial manipulators is typically applied in path planning. In particular, the path is typically planned in configuration space instead of work space for industrial manipulator. The dynamic and kinematic constraints of an industrial manipulator make the shape of configuration space different from the work space. However, the configuration space and work space typically turn out to be same for mobile robots. For instance, the path planning of circular-shaped robot considers size of the obstacle grown by the diameter of the circular body of the robot. This way the dynamic and kinematic constraints are taken care by not changing the shape of configuration space from work space.

There exist largely two popular classes of path planning: graph search and potential-field-based techniques. In particular, almost all the path-planning techniques use continuous or discrete map representation. A potential field approach is based on applying a mathematical function on the free space. The gradient outcome of this function gives the instantaneous direction toward the goal. On the contrary, graph-based approach first constructs a graph-based representation containing edges and nodes. Then the shortest path is computed based on various graph search approaches.

3.4.1 Potential Field Path Planning

In potential field path planning approach of Hwang et al. (1992), Wang and Chirikjian (2000), Rasekhipour et al. (2016), and Lee and Park (2003), the robot is directed toward the goal position by considering an artificial potential field. The potential field due to goal position acts as an attractive force, whereas potential field due to obstacles acts as repulsive force on robot. The robot experiences a superposition of all attractive and repulsive forces and the final movement is in the direction of the resulting force. In addition to path planning, the potential field also acts as a control law for mobile robots, which can provide the next control action to be performed. The basic potential field approach neglects the robot's orientation and the resultant potential field is only in 2D.

The attractive potential field is typically considered as a function of Euclidean distance between the robot location (x, y) and the goal location given (x_{goal}, y_{goal}) by,

$$U_{att} = \frac{1}{2} \cdot k_{att} \cdot d_{goal}^2 \tag{3.8}$$

where $d_{goal} = \sqrt{(x - x_{goal})^2 + (y - y_{goal})^2}$.

The attractive force as illustrated in Figure 3.6 is computed by differentiating U_{att},

$$F_{att} = -\Delta U_{att}$$

which tends to 0 as the robot reaches the goal.

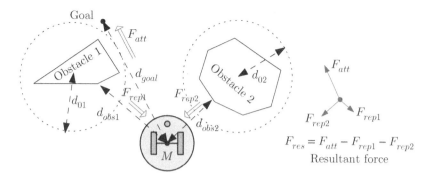

Figure 3.6 Illustration of potential function technique

The repulsive function needs to be inversely proportional to the distance with the obstacle, because the repulsive force should be high near the obstacle. Moreover, it should not influence when the robot is far away from the obstacle. Therefore, a typical repulsive potential field is given by

$$U_{rep} = \begin{cases} \frac{1}{2} \cdot k_{rep} \cdot \left(\frac{1}{d_{obs}} - \frac{1}{d_0} \right) & \text{if } d_{obs} \leq d_0 \\ 0 & \text{if } d_{obs} > d_0 \end{cases} \tag{3.9}$$

where d_0 is the minimum distance from the obstacle after which the repulsive force is considered negligible and d_{obs} is the distance of the robot from the obstacle as illustrated in Figure 3.6. Figure 3.6 shows Obstacles 1 and 2 where d_{obs1} and d_{obs2} are the distances of the robot to the obstacle and d_{goal} is the distance of the robot to the goal. The distance d_{01} and d_{02} are minimum distances corresponding to the Obstacles 1 and 2 after which the effect of repulsive force is negligible. The repulsive force is obtained in similar manner as attractive force. The control law is a combination of attractive and repulsive forces with an appropriate selections of their weights through the constants k_{att} and k_{rep}. Hence, the control law can be developed as follows:

$$u = -\Delta U_{att} - \Delta U_{rep} \tag{3.10}$$

The control law given by (3.10) works well in the presence of one obstacle as there exists only one repulsive force vector influencing the control law. However, when there are multiple obstacles around the mobile robot, the influence of multiple repulsive functions can either introduce local minima or generate oscillations in subsequent commands. The potential field approach is known for its limitations of having local minima and issues in dealing with non-convex objects. These can lead to robot getting stuck at local minima or oscillating about a certain position. However, potential field technique is still a popular choice for its simplistic approach and minimal sensing requirements.

Figure 3.7 Improvement in potential field technique illustrating the closest distance direction nearby obstacle orthogonal to the robot's forward direction

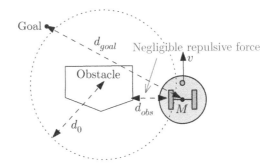

To overcome the limitations of potential function technique, there are several improvements that have been made such as considering the orientation of the robot. If the closest distance direction of the nearby obstacle is orthogonal to the robot's forward direction, then its effect is accounted negligible as shown in Figure 3.7. In other words, the repulsive force is computed based on the distance to a nearby obstacle when the obstacle is in the direction of motion.

3.4.2 Graph-based Path Planning

In graph-based path planning, a connectivity graph representing the free space of the environment is developed and referred for computation of a shortest path. Usually, the construction of graph is performed online. These are connectivity graphs consisting of nodes and edges as illustrated in Figure 3.8. There are various techniques used for graph construction such as visibility graph, Voronoi diagram, and cell decomposition methods. These techniques are not discussed in this text, but a good context about them can be referred to in Siegwart et al. (2011). In this section, we will discuss three most popular deterministic graph section techniques: Dijkstra's algorithm, A* algorithm, and RRT (Rapidly-exploring Random Trees algorithm).

Figure 3.8 Connectivity graph

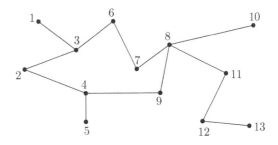

3.4.2.1 Dijkstra's Algorithm

This algorithm, seen in Wang et al. (2011), Zhang and Zhao (2014), Tan et al. (2006), and Dong et al. (2010), finds the shortest path between start node and end node based on breadth-first search. The iterative algorithm begins the search from start node and explores all the neighboring nodes. The edges between two nodes have a finite cost which represents the cost of travel. Then, the neighboring nodes are further explored till the destination node is reached. Beginning from the start node as the current node, it is marked as *active* and the neighboring nodes are included in an *open set*. In each iteration, an active node is selected. The costs of all the neighbors of the active node are assigned, and the current node is marked as *visited* and removed from the *open set*. The active node in the next iteration is selected is one among the *open set* having same cost as the current active node. If no such node with the same cost is available, then the next highest cost node is selected as the current node. Figures 3.9 and 3.10 illustrate the cost computation of cell adjacent/diagonal to the one with cost same after each iteration using Dijkstra's algorithm where map is in cellular decomposition form. The cost of movement (g) to move in adjacent cell is 1 unit and adjacent diagonal cell is 2 unit. The path is planned from S start cell to G goal cell. The cells in the *open set* are in light gray color and cells in *visited* set are in dark gray color. Once the cost is assigned to the goal node, the algorithm backtracks from the goal node and finds path to the source node that turns out to be the shortest path between the start and goal nodes.

3.4.2.2 A* Algorithm

This algorithm is similar to Dijkstra's algorithm with an addition of an extra heuristic function to the cost function. This heuristic function includes the distance between any node to the goal. This distance can be in terms of Euclidean distance or the direct distance between the current node and goal node without obstacles in case of cell-based maps. The inclusion of heuristic in the cost function dramatically reduces the number of node expansion and finds the goal location faster than Dijkstra's algorithm. Similar to the Dijkstra's algorithm, the A* algorithm seen in Candra et al. (2020), Seo et al. (2009), Le et al. (2018), and Bell (2009), begins the search by expanding the start node and considering all its neighbors in the *open set*. The node in the open set which has lowest cost function value is selected for the next expansion. This continues till the goal location is reached. The shortest path, thus, can be back-traced from the goal location. An illustration of the algorithm is given in Figure 3.11. The cost function is given by ($g + h$) value where g is the cost of the cell from start (like the one in the Dijkstra's algorithm) and h is the heuristic cost which is the distance from the goal. The cells in the *open set* are in light gray color and cells in *visited* set are in dark gray color. We can notice that the number of visited cells obtained from A* algorithm is less as compared to that obtained through Dijkstra's algorithm.

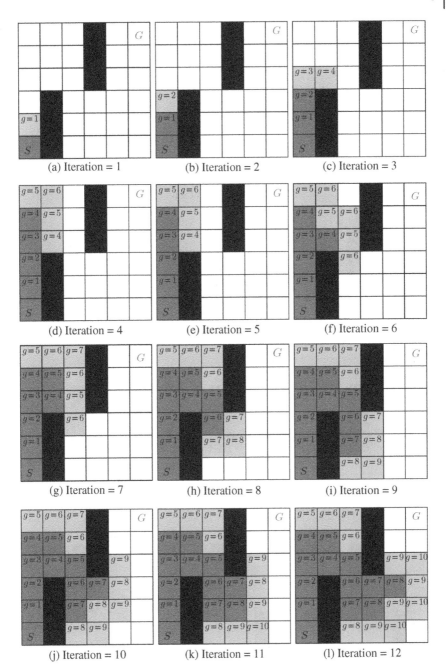

Figure 3.9 Illustration of cost computation of each cell using Dijkastra's algorithm

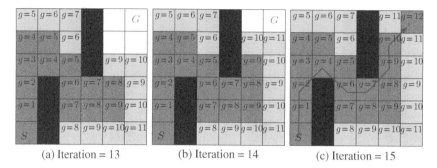

(a) Iteration = 13   (b) Iteration = 14   (c) Iteration = 15

Figure 3.10 Illustration (continued) of cost computation of each cell using Dijkastra's algorithm

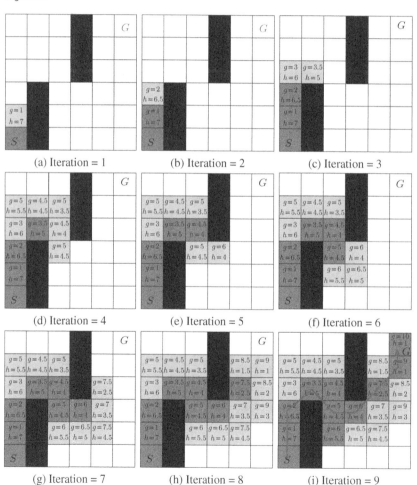

(a) Iteration = 1   (b) Iteration = 2   (c) Iteration = 3

(d) Iteration = 4   (e) Iteration = 5   (f) Iteration = 6

(g) Iteration = 7   (h) Iteration = 8   (i) Iteration = 9

Figure 3.11 An illustration of iterations in the A* algorithm

Figure 3.12 RRT algorithm

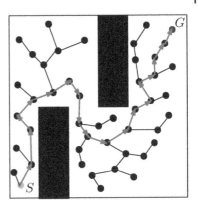

3.4.2.3 Rapidly-exploring Random Trees (RRT) Algorithm

The evolution of RRT in Kuffner and LaValle (2000), Karaman et al. (2011), Kuwata et al. (2008), Islam et al. (2012), Salzman and Halperin (2016), and Urmson and Simmons (2003) grows in several directions while exploring the environment. The idea is to populate nodes in free space of the environment that can form a connected tree between the source and goal nodes. The algorithm begins with an initial node and successively adding more nodes, by connecting edges. During each step, a random node q_{rand} configuration is selected in the free space. The tree node q_{near} which is closest to q_{rand} is selected. An edge connecting q_{rand} and q_{near} is found non-intersecting with the obstacles, the complete edge then represents collision-free movement on it and therefore the configuration q_{near} and the edge between q_{rand} and q_{near} are added to the tree. Thus, this algorithm generates the tree online, until the goal node is reached. The shortest path is thus computed between the start the node and goal node using graph search algorithm as shown in Figure 3.12. The RRT algorithm typically speeds up by growing the tree simultaneously from the start and the goal node, leading to faster convergence.

The extensions of techniques developed for a mobile robot are investigated in multi-agent scenarios. The Section 3.5 describes additional challenges to be addressed in embedded implementation when multiple mobile robots are collectively performing a common goal.

3.5 Multi-agent Scenarios

Many robotic applications emerged that explore the possibility of achieving a common objective with a team of robots. The team of robots or multiple agents perform a given task in a collaborative and coordinated manner. These collaborative agents/robots agree to follow a common strategy that avoids redundancy and work toward an optimal objective. Typically, these robots are small in size and resource constrained so that they show the efficacy of achieving an objective compared to that achieved using single agent.

Multi-agent scenarios in the context of mobile robotic applications perform a common objective. For example,

- in the swarm aggregation, multiple agents (hundreds in number) move toward a common goal location
- multi-agent patrolling where two or more agents perform inspection of given locations repeatedly
- multi-agent mapping where all parts of an area is to be covered optimally to prepare a representation of an environment using multiple agents/robots
- in master–slave configuration, robots copies the actions of master robot
- in the leader–follower configuration, the follower copies the movement of leader robot
- and many more.

While the common objectives in these or emerging multi-agent mobile robotic applications are different each of these problems needs to address primitive problems that are common. The typical primitive problems are the following:

- *Inter-agent communication*: This defines the way agents communicate to each other. It can be one-to-one or one-to-many and maintain a communication network so that each agent is always a part of this network. In particular, it is important to ensure that all agents are connected at all times through a communication network or at least their information is available in the network in some or the other form. The information exchange between agents is specific to the application.
- *Control methodology*: While the agents are independently executing the commands in the multi-agent applications, the decision-making commands are generated through either a centralized station (cloud computing) or independently (edge computing). The information exchange and decision making are dependent on each other. Hence, control methodology addressing inter-agent communication providing feedback and controller design making decisions is to be formulated. Since the controller is implemented in an embedded platform for a small-size mobile robot in the multi-agent application, it is necessary to implement embedded controller design.
- *Multi-agent collision avoidance*: A few popular collision detection and avoidance techniques for a mobile robot to operate in the environment populated with static and dynamic obstacle is discussed in Section 3.2. In the multi-agent scenario, the inter-agent collision avoidance can be handled in a specific way as the agent dynamics and information (current state) are available to the system. The collision avoidance strategy can be independently handled by the

Figure 3.13   Inter-agent collision detection

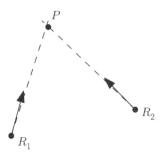

decision-making mobile robot, but this would not ensure avoidance with other robots who are also handling the avoidance independently. More specifically, the independent decision making may not ensure collision avoidance if not taken care in a collaborative manner. Furthermore, the benefits of cooperation can be explored to avoid redundancy in decisions and guarantee collision avoidance. For example, let the agents R_1 and R_2 be moving with the velocities v_1 and v_2, respectively, as shown in Figure 3.13. These agents are following independent collision avoidance strategy when collision is detected at point P by each of them. The independent decision to avoid collision will not ensure avoidance if each robot takes decision of slowing down the speed by the same amount. Moreover, each robot is attempting to avoid collision and adding redundancy in the inter-agent collision strategy.

There are two ways to handle the inter-agent collision in a collaborative multi-agent scenario. (i) The inter-agent collision avoidance strategy can be designed at the system level that either ensures no collision while designing motion planners or decision is taken at a centralized station (cloud or master robot). (ii) A cooperative strategy such as assigning priority for independent robots be agreed among the robots. The robot with higher priority among the ones in the collision may take a call on avoiding the collision. For embedded implementation in the individual robot, it acts as a condition whether to act on collision avoidance or not. However, other interfaces are needed on-board that give feedback on the engaged agent when collision is detected. Therefore, the feedback (or relevant information extraction from on-board sensors) is crucial as compared to addressing control objectives.

- *Multi-agent localization*: Each agent in a multi-agent scenario can independently solve the localization problem and locate itself. However, each agent can further improvise its location by communicating with other agents and benefit from the collaboration of multiple agents. In certain scenarios, an extra step is recommended to calibrate the location of multiple agents as the location (state) estimation of each agent as well as with respect to each other decides the outcome of collaboration. At the embedded implementation level, it the step toward improving state estimates and perhaps implemented by adding states of other agents in the multi-agent system in the state vector of an agent. The dimension of system at the agent level increases tremendously and would involve higher computing resources that is to be matched with the available resources on the selected embedded platform.

3.6 Notes and Further Readings

The approaches described in this chapter for obstacle avoidance are the most popular and basic ones. However, there are many other approaches developed for obstacle avoidance, such as Nearness Diagram (ND), which is an improvement toward VFH where it takes into account more precise geometric, kinematic, and dynamic constraints of Minguez and Montano (2000, 2004), Durham and Bullo (2008), and Minguez et al. (2001). A gradient based technique of Konolige (2000) and Farinelli and Iocchi (2003) is a generalization of wavelet propagation techniques and generates a gradient direction at a given time instant for any grid. Some techniques, such as bubble-band technique of Quinlan and Khatib (1993) and Susnea et al. (2010), need prior knowledge of the entire environment. Fuzzy and neuro-fuzzy approaches have also been used for obstacle avoidance as seen in Tzafestas and Tzafestas (1999).

On state estimation for localization, many probabilistic localization techniques such as Kalman Filter localization, Unscented Kalman Filter localization, and Monte Carlo localization are also popular techniques that are used in various commercial and research platforms. Interested readers can refer to works reported in Chen (2011), Fuentes-Pacheco et al. (2015), and Panchpor et al. (2018)

Interesting works on multi-agent scenarios, where motivation comes from the self-organizing behavior in biological beings, have been reported. Interested readers may refer to works in Sahin (2004) and Brambilla et al. (2013) for modeling collective behaviors with limited sensing and communication resources, Parker and Zhang (2008) for task sequencing and mission control, or Moeslinger et al. (2011), Balch and Hybinette (2000), and Chiew et al. (2015) for decentralized flocking and coordination. The miniaturization of robots specifically for using them in multi-agent scenarios can be referred to robots like the ones reported in Alice Caprari et al. (2001), Kilobots Rubenstein et al. (2012), GRITSBots Pickem et al. (2015), etc.

4

Bottom-up Method

In many robotic application settings, the choice of embedded platform is typically not available in the following two scenarios: (i) The embedded platform is selected based on the requirements of sensing, communication, power, operating conditions, and similar parameters. (ii) The controller design or planning exercise is an up gradation procedure. In such scenarios, the design procedure must consider the limitations of an embedded processor.

In order to address the limitations of computing digitally, two generic practices are discussed in this chapter. Firstly, we discuss a popular algorithm, CORDIC (COordinate Rotation DIgital Computer), which is already used in many applications. This algorithm and its variants can compute logarithmic, trigonometric, exponential, and many such computations using only addition/subtraction and shifting the bits. Adder/Subtractor (AS) and Shifters are building blocks of any embedded platform. Hence, the use of CORDIC for such computations is possible in almost every embedded platform. Because of its capabilities, the CORDIC as a separate block is now available in most DSP (Digital Signal Processor), microcontroller, and FPGA (Field Programmable Gate Array) platforms. This chapter presents the concept of CORDIC and its use in computing typical computations required in controller design. Next, this chapter discusses a generic method of designing robotic tasks using interval analysis. The interval analysis, here, is used as a basis for designing algorithms for robotic applications. This approach is particularly suitable for embedded implementations because the interval arithmetic and algorithms involve additions and multiplication/division by 2. The multiplication and division by 2 are nothing but shifting left and shifting right in the digital operations respectively.

Embedded Control for Mobile Robotic Applications, First Edition.
Leena Vachhani, Pranjal Vyas, and Arunkumar G. K.
© 2022 The Institute of Electrical and Electronics Engineers, Inc. Published 2022 by John Wiley & Sons, Inc.
Companion website: www.wiley.com/go/vachhani/embeddedcontrolforroboticapp

4.1   Computations Using CORDIC[1]

The transformation matrix $R(\phi)$ for rotating the vector (x, y) by a counter clockwise angle ϕ is given by (3.5) and revisited here.

$$R(\phi) = \begin{bmatrix} \cos(\phi) & -\sin(\phi) \\ \sin(\phi) & \cos(\phi) \end{bmatrix}$$

Let the angle of rotation ϕ be expressed for q iterations as $\phi = \sum_{i=0}^{q-1} d_i \phi_i$, where $d_i = +1$ for counter clockwise rotation and $d_i = -1$ for clockwise rotation, or

$$d_i = \begin{cases} +1 & \text{for counter clockwise rotation} \\ -1 & \text{for clockwise rotation} \end{cases}$$

The sequence of micro-rotations by angles $d_i \phi_i$ for each $i = 0, 1, 2, \ldots, q-1$ is $R(\sum_{i=0}^{q-1} d_i \phi_i) = \prod_{i=0}^{q-1} R(d_i \phi_i)$. Therefore, the resulting vector is given by

$$\begin{bmatrix} x_f \\ y_f \end{bmatrix} = \prod_{i=0}^{q-1} R(d_i \phi_i) \begin{bmatrix} x \\ y \end{bmatrix} \text{ or}$$

$$\begin{bmatrix} x_f \\ y_f \end{bmatrix} = \prod_{i=0}^{q-1} \left(\cos(\phi_i) \begin{bmatrix} 1 & -\tan(\phi_i) \\ \tan(\phi_i) & 1 \end{bmatrix} \right) \begin{bmatrix} x \\ y \end{bmatrix}$$

The CORDIC (see Volder (1959)) is an iterative algorithm that computes the rotation of vector (x, y) by the counter clockwise angle ϕ using sequence of micro-rotations by angles $\phi_i = \arctan(2^{-i})$ for each i. This choice of angles renders iterative computations merely limited to shift and add operations. Following are the iterative equations for the CORDIC algorithm:

$$x_{i+1} = x_i - d_i 2^{-i} y_i$$
$$y_{i+1} = y_i + d_i 2^{-i} x_i$$
$$\phi_{i+1} = \phi_i - d_i \arctan\left(2^{-i}\right) \tag{4.1}$$

for each $i = 0, 1, \ldots, q-1$, and the resulting vector is computed using a multiplier by $[x_f \ y_f]^T = K_q [x \ y]^T$, where CORDIC constant $K_q = \prod_{i=0}^{q-1} \cos(\arctan(2^{-i}))$. The CORDIC constant asymptotically reaches 0.607252935 .... It is worth noting that $K_q \approx 0.6073$ for $q = 6$ itself. Since the constant multiplier can also be implemented using shift and add operations, the CORDIC algorithm is a popular choice for computing vector rotation in embedded platforms. There are two modes of operation for CORDIC, namely, rotation and vectoring. In the rotation mode, the vector is rotated by the desired angle, whereas the vector is rotated to align with the X-axis in the vectoring mode. The result of vectoring mode is the vector

---

1 Modified from [1].

length and angle made by the input (initial) vector with X-axis. The direction of rotation is given by

$$d_i = \begin{cases} +1 & \text{if } \phi_i \geq 0 \text{ for rotation mode} \\ & \text{or } y_i \leq 0 \text{ for vectoring mode} \\ -1 & \text{otherwise} \end{cases} \quad (4.2)$$

The number of iterations required for q-bit word length by the conventional CORDIC is q. The architecture for conventional CORDIC is shown in Figure 4.1. It comprises Barrel Shifter (BS), Adder/Subtractor (AS), and Memory (M). The critical path as computed in Meher and Park (2019) for x and y data paths has BS and AS elements, whereas it has M and AS elements for ϕ data path. Given the propagation delays in BS, AS, and M blocks as T_{BS}, T_{AS}, and T_M, respectively, the delay in

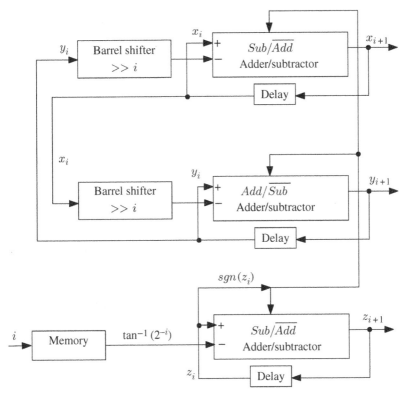

Figure 4.1 Architecture of conventional CORDIC.

critical path in conventional CORDIC is given by

$$T_{conv} = \begin{cases} T_{BS} + T_{AS} & \text{if } T_{BS} > T_M \\ T_M + T_{AS} & \text{otherwise} \end{cases} \tag{4.3}$$

There are many matrix and vector computations possible using CORDIC. There are many ways by which computations of algebraic functions using CORDIC has been worked out. This has been possible by using various combinations of initializing the vector in vectoring and rotation modes. We will now look into the way vectors are initialized for Coordinate transformation and computation of exponential functions using CORDIC. For computations of other functions, readers are requested to refer to further readings.

4.1.1 Coordinate Transformation

As discussed in Section 3.1, there are many instances when the rotation of vector is computed. While computation of rotation matrix using CORDIC is straightforward (refer (4.1)), the coordinate transformation from Cartesian to polar and vice versa is also possible using CORDIC algorithm. Let the Cartesian coordinates of a point P be (x, y) and its polar coordinates be (r, θ). Now, let's look into the way by which the CORDIC algorithm is used for coordinate transformation.

4.1.1.1 Cartesian to Polar Coordinates Conversion

The CORDIC algorithm in vectoring mode rotates the vector (x_0, y_0) to (x_f, y_f) such that the final vector (x_f, y_f) is close to X-axis ($y_f \to 0$) after q iterations. The amount of rotation recorded in ϕ_q is nothing but the angle θ. The length of the vector (x_f, y_f) is x_f as $y_f \approx 0$. After the CORDIC correction, we get the length r. Hence, the outcome of CORDIC in vectoring mode is $(r, \theta) = (x_f/K_q, \phi_q)$.

4.1.1.2 Polar to Cartesian Coordinate Conversion

Now, the rotation mode of CORDIC algorithm facilitates polar to Cartesian coordinate conversion by initializing the vector (x_0, y_0) to $(r, 0)$ with $\phi_0 = \theta$. Note that the initial vector is now along the X-axis. In the rotation mode, the vector is rotated to reach a vector that makes an angle of ϕ_0 with respect to the X-axis. Hence, the final vector (x_f, y_f) is the Cartesian coordinates of the point P scaled by CORDIC constant. In particular, $(x, y) = (x_q/K_q, y_q/K_q)$ with $\phi_q \approx 0$.

4.1.2 Exponential and Logarithmic Functions

For computing the exponential and logarithmic functions, the hyperbolic functions are used. Let the exponent of an angle θ be t, or $e^\theta = p$. We know that

$$p = e^\theta = \sinh \theta + \cosh \theta$$

Similarly,

$$\theta = \ln p = 2\tanh^{-1}\left(\frac{y}{x}\right), \text{where } y = \theta - 1 \text{ and } x = \theta + 1$$

In order to compute hyperbolic functions, the CORDIC algorithm is executed by modifying (4.1) by the following iterative equations:

$$x_{i+1} = x_i - d_i 2^{-i} y_i$$
$$y_{i+1} = y_i + d_i 2^{-i} x_i$$
$$\phi_{i+1} = \phi_i - d_i \tanh^{-1}\left(2^{-i}\right) \quad\quad\quad (4.4)$$

Notice the change in iterative equation of ϕ_i, the arctan($\cdot$) is replaced with arctan($\cdot$). Accordingly, the CORDIC constant K_q is given by

$$K_q = \prod_{i=0}^{q-1} \cos(\tanh^{-1}(2^{-i}))$$

Now, in order to compute e^θ, the rotation mode of CORDIC is used with initial vector at $(1,0)$ and angle $\phi_0 = \theta$. This way the final outcome is obtained in $x_f = \cosh(\theta)/K_q$ and $y_f = \sinh\theta/K_q$. And now, $e^\theta = (x_f + y_f)/K_q$.

While computation of exponential uses methodology same as Polar to Cartesian coordinates conversion, the computation of logarithm uses methodology of Cartesian to Polar coordinates conversion. For computing $\ln t$, the vectoring mode of CORDIC is used to compute logarithm by initializing $x_0 = p + 1$ and $y_0 = p - 1$. The obtained final vector is of the form $(x_f, y_f) = (\ln p/2K_q, 0)$ as the outcome of CORDIC is $(\tanh^{-1} y_0/x_0, 0)$. As a result, we also get $\phi_q = \sqrt{x_0^2 - y_0^2}$ which can be used for computing square root.

4.2 Interval Arithmetic[2]

In the bottom-up approach, it is important to note that the computations and algorithms are implemented in a particular embedded platform. A generic approach where the algorithms can be developed regardless of the choice of embedded platform is presented followed by its specific implementation for FPGA is discussed.

4.2.1 Basics of Interval Arithmetic

This section covers a brief introduction of the interval arithmetic used in this chapter. The interval consists of a range of values which describe probable values

---

2 Taken from Vyas (2017).

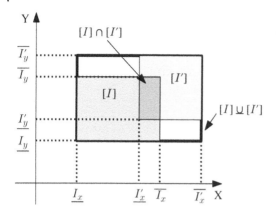

Figure 4.2 Interval intersection and interval hull

for a variable. Let the set of real numbers be $\mathbb{R}$. A closed interval $[I]$ for a set of real numbers ranging from a to b is given by

$$[a, b] = x \in \mathbb{R} : a \leq x \leq b \qquad (4.5)$$

The infimum and supremum of an interval $[I]$ are denoted by I and $\bar{I}$. In particular, (4.5) is defined as $[I] = [I, \bar{I}]$ where $I = a$ and $\bar{I} = b$. Two intervals $[I]$ and $[I']$ are said to be equal if they are the same sets. An interval $[I]$ is called degenerate if $I = \bar{I}$.

The *intersection* of two intervals $[I]$ and $[I']$ is an interval and is defined as

$$[I] \cap [I'] = z : z \in [I] \text{ and } z \in [I']$$

$$= [\max \{I, I'\}, \min \{\bar{I}, \bar{I'}\}]. \qquad (4.6)$$

In general, the *union* of two intervals $[I] \cup [I']$ does not form an interval. However, the *interval hull* of two intervals $[I]$ and $[I']$, defined by,

$$[I] \underline{\cup} [I'] = z : z \in [I] \text{ or } z \in [I']$$

$$= [\min \{I, I'\}, \max \{\bar{I}, \bar{I'}\}] \qquad (4.7)$$

is always an interval since it is an over-approximation and can be used in interval computations. Thus, we have $[I] \cup [I'] \subseteq [I] \underline{\cup} [I']$ for any two intervals $[I]$ and $[I']$. The *intersection* and *interval hull* of two intervals $[I]$ and $[I']$ for the 2D intervals is shown in Figure 4.2. The interval hull is a very useful operation in interval analysis. If the contents of two intervals form a result, then the interval hull will also contain the result.

A few useful terms in interval arithmetic are:

- The width of an interval $[I]$ is defined as $w([I]) = \bar{I} - I'$.
- The absolute value of an interval $[I]$, denoted by $|I|$ is defined as $|I| = \max\{|I|, |\bar{I}|\}$.
- The midpoint of $[I]$ is given by, $m([I]) = \frac{1}{2}\left(I + \bar{I}\right)$.
- $[I] < [I']$ means that $\bar{I} < I'$.
- $[I] \subseteq [I']$ if and only if $I' \leq I$ and $\bar{I} \leq \bar{I}'$.
- Any interval $[I]$ can be expressed as

$$[I] = m([I]) + \left[-\frac{1}{2}w([I]), \frac{1}{2}w([I])\right]$$
$$= m([I]) + \frac{1}{2}[-w([I]), w([I])]$$

- An n-D vector $[I] \in \mathbb{IR}^n$ with interval components is described as

$$[I] = [I_1] \times [I_2] \times \cdots \times [I_{n-1}] \times [I_n] \tag{4.8}$$

where $[I_i]$ is the projection of $[I]$ onto ith axis for $i = 1, 2, \ldots, n$ and $\mathbb{IR}$ is the set of all intervals of real numbers.
- The *intersection* of two interval vectors $[I] \cap [I'] = \left([I_1] \cap [I'_1], \ldots, [I_n] \cap [I'_n]\right)$.
- The width of an interval vector $[I] = \left([I_1], \ldots, [I_n]\right)$ is the largest of the widths of any of its component intervals.
- The midpoint of an interval vector $[I] = \left([I_1], \ldots, [I_n]\right)$ is $m([I]) = (m([I_1]), \ldots, m([I_n]))$.
- The norm of an interval vector $[I] = \left([I_1], \ldots, [I_n]\right)$ is the generalization of absolute value given by $\| I \| = \max |I_i|$.
- *Addition*: $[I] + [I'] = [I + I', \bar{I} + \bar{I}']$.
- *Subtraction*: $[I] - [I'] = [I - \bar{I}', \bar{I} - I']$.
- *Multiplication*: $[I] * [I'] = [\min\{I \cdot I', I \cdot \bar{I}', \bar{I} \cdot I', \bar{I} \cdot \bar{I}'\}, \max\{I \cdot I', I \cdot \bar{I}', \bar{I} \cdot I', \bar{I} \cdot \bar{I}'\}]$
- *Division*: $[I]/[I'] = [I] * (1/[I'])$ where $1/[I'] = [1/\bar{I}', 1/I']$ provided $0 \notin [I']$.

4.2.2 Inclusion Function and Inclusion Tests

Consider a function $f : \mathbb{R}^n \mapsto \mathbb{R}^m$. The interval function $[f] : \mathbb{IR}^n \mapsto \mathbb{IR}^m$ is an *inclusion function* for f if any $[I] \in \mathbb{IR}^n, f([I]) \subset [f]([I])$. For instance, consider a function $f : \mathbb{R}^2 \mapsto \mathbb{R}^2$ described for variables i_1 and i_2. Let the intervals for i_1 and i_2 be given by $[I_1]$ and $[I_2]$. The image of function $f([I])$ can have any shape. The inclusion function $[f]$ of f computes an interval which encloses $f([I])$ and guarantees to contain it as shown in Figure 4.3.

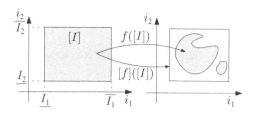

Figure 4.3 Inclusion function (Jaulin et al., 2001)

Similarly, an *inclusion test* can be used to prove that all the points in given interval satisfy a given property, or to prove that none of them does. These tests involve Boolean intervals, which will be presented first. A Boolean number is an element of the set $\mathbb{B}$ defined as

$$\mathbb{B} \triangleq \{false, true\} = \{0, 1\}$$

Therefore, a *Boolean interval* $\mathbb{IB}$ is described by

$$\mathbb{IB} = \{\emptyset, 0, 1, [0, 1]\}$$

where $\emptyset$ stands for *impossible*, 0 for *false*, 1 for *true*, and $[0, 1]$ for *partial*, where few points of the interval result in true while the remaining result in false. A *test* is a function $t : \mathbb{R}^n \mapsto \mathbb{B}$ for $[I] \in \mathbb{IR}^n$ given by

$$t_I(p) = \begin{cases} 1 & \text{if } p \in [I] \\ 0 & \text{otherwise(partial inclusion of } [I'] \text{ in } [I]) \end{cases} \tag{4.9}$$

An *inclusion test* $[t_I] : \mathbb{IR}^n \mapsto \mathbb{IB}$ is an inclusion function for t_I such that for any $[I'] \in \mathbb{IR}^n$, $[t_I]([I'])$ satisfies

$$[t_I]([I']) = \begin{cases} 1 & \text{if } t_I(p) = 1 \ \forall \ p \in [I'] \\ 0 & \text{if } t_I(p) = 0 \ \forall \ p \in [I'] \\ [0, 1] & \text{otherwise} \end{cases}$$

If the inclusion test is *true* ($t_I([I]) = 1$), i.e. $[I'] \subset [I]$, each element in $[I']$ satisfies the inclusion test. If the inclusion test result is *indeterminate* ($t_I([I]) = [0, 1]$), then there is partial inclusion of $[I']$ in $[I]$, i.e. not all elements of $[I']$ are contained in $[I]$. An *XNOR* operation $\odot$ between two Booleans Inp_1 and Inp_2 is described using Table 4.1. Table 4.1 also presents the output of $Inp_1 \odot Inp_2$ for each combination of inputs. In particular, the symbol X corresponds to "don't care." The notation $\odot_{i=1}^{n} Inp_i$ describes the *XNOR* operation for n inputs Inp_i for $i = 1, 2, \ldots, n$.

Table 4.1   Truth table for *XNOR*

Inp_1	Inp_2	$Inp_1 \odot Inp_2$
0	0	1
1	0	0
0	1	0
1	1	1
X	$[0,1]$	$[0,1]$
$[0,1]$	X	$[0,1]$

4.3   Collision Detection Using Interval Technique[3]

As we learnt in Chapter 3 (refer Section 3.2), collision detection using local sensing methods is challenging and is possible to achieve a solution using various sensors. The choice of sensor depends on the problem being addressed. In particular, the collision detection and avoidance algorithm decides if a particular sensor provides the necessary data as input and vice versa. Furthermore, the decision taken by the collision avoidance algorithm is typically an independent choice of the mobile robot. Therefore, it is challenging to ensure collision detection and avoidance in multi-robot scenarios and/or dynamic environments where robots or dynamic obstacles do not convey the decision to avoid collision.

When we try solving the collision avoidance problem targeting implementation and addressing the challenge of independent decision making, we can present the problem in terms of interval sensors. Presenting the problem as interval analysis problem can then provide collision-free intervals that guarantee collision avoidance with any decision taken by moving robot/obstacle in a given horizon. Let us formulate the interval-based problem and assume that the maximum velocity change of the decision-making robot as well as the dynamic obstacles are known.

Let the pose of robot and dynamic obstacle be represented by polar coordinates (r, θ), where $r \in \mathbb{R}$ and $\theta \in \mathbb{S}^1 = [0, 2\pi)$. The set of positions of a robot which can be achieved in the time horizon within the velocity bounds is expressed in terms of polar intervals. Let the instantaneous poses of decision-making robot and dynamic obstacle be $O(x,y,\theta)$ and $O'(x',y',\theta')$, respectively. Let the corresponding maximum linear and angular velocities be $[v_{min}, v_{max}, \omega_{min}, \omega_{max}]$

---

3 Taken from Vyas (2017)

and $[v'_{\min}, v'_{\max}, \omega'_{\min}, \omega'_{\max}]$. Now, the two intervals $[I]$ and $[I']$ corresponding to all possible positions of the robot and dynamic obstacle, respectively, in time horizon ΔT are given by

$$[I] = [(v - v_{\min})\Delta T, (v + v_{\max})\Delta T]$$
$$\times [\theta - \omega_{\min}\Delta T, \theta + \omega_{\max}\Delta T]$$
$$[I'] = [(v' - v'_{\min})\Delta T, (v' + v'_{\max})\Delta T]$$
$$\times [\theta' - \omega'_{\min}\Delta T, \theta' + \omega'_{\max}\Delta T]$$

Figure 4.4 illustrates intervals $[I]$ and $[I']$ corresponding to the possible positions of the robot and dynamic obstacle in ΔT time. The intervals $[I]$ and $[I']$ are shown in light gray color and their intersection in interval representation $[J]$ with respect to the decision-making robot in dark gray color. The collision between the robot and dynamic obstacle is detected by finding the possibility of inclusion of $[I]$ in $[I']$ and vice versa. The corresponding collision-free interval of the decision-making robot is computed for safe maneuvering of robot. The robot can reach to any point in the collision-free intervals in the next ΔT time ensuring collision avoidance. The technique to find the collision-free intervals is described in Section 4.4. The interval corresponding to the robot describes the possible moves of the robot in a given time interval. The shape and size of the interval are determined by the maximum possible changes in linear and angular velocities in this interval-based problem formulation. For example, the interval for an omnidirectional robot is the entire

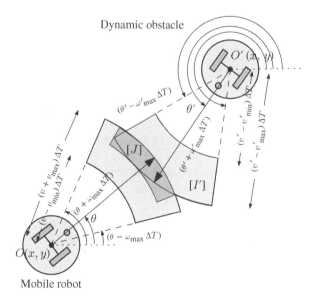

Figure 4.4 Colliding objects: the decision-making robot and dynamic obstacle

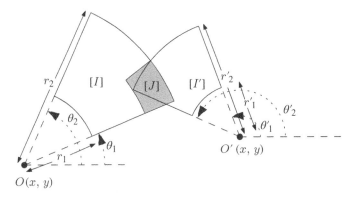

Figure 4.5 Interval representation of velocities of decision-making robot that may result in collision

Table 4.2 Interval notations for sub-intervals

Curves	Interval notation	Description
Arc γ_1	$[\gamma_1]$	$[r_1, r_1] \times [\theta_1, \theta_2]$
Arc γ_2	$[\gamma_2]$	$[r_2, r_2] \times [\theta_1, \theta_2]$
Arc γ_1'	$[\gamma_1']$	$r_1', r_1'] \times [\theta_1', \theta_2']$
Arc γ_2'	$[\gamma_2']$	$[r_2', r_2'] \times [\theta_1', \theta_2']$
Circle η	$[\eta]$	$[r_2, r_2] \times [0, 2\pi]$
Circle η'	$[\eta']$	$[r_2', r_2'] \times [0, 2\pi]$
Line-segment l_1	$[l_1]$	$[r_1, r_2] \times [\theta_1, \theta_1]$
Line-segment l_2	$[l_2]$	$[r_1, r_2] \times [\theta_2, \theta_2]$
Line-segment l_1'	$[l_1']$	$[r_1', r_2'] \times [\theta_1', \theta_1']$
Line-segment l_2'	$[l_2']$	$[r_1', r_2'] \times [\theta_2', \theta_2']$

circle centered at the current position of the robot. Hence, collision of an omnidirectional robot is found by considering $[I] = [0, \sqrt{v^2 + v_n^2}\Delta T] \times [0, 2\pi]$, where v and v_n are the velocities along the forward and its normal directions, respectively. The proposed technique for collision detection computes the inclusion of $[I]$ in $[I']$ or vice versa using the inclusion test $t[I][I']$ described in Section 4.2.2. In order to find out the inclusion of $[I]$ in $[I']$, it is sufficient to compute the inclusion for individual boundary curves. The intervals $[I]$ and $[I']$ consist of boundary curves as arcs and line-segments as illustrated in Figure 4.5 which can individually be represented in interval notations. Table 4.2 defines various sub-interval notations for $[I]$ and $[I']$ used in this chapter.

The steps to compute $[t_{l'}]([I])$ are described in Algorithm 1. The inclusion of $[I]$ in $[I']$ is computed using the inclusion of individual boundary curves intervals of $[I]$ and $[I']$. These boundary curves for the illustrative case are listed as $[l_i]$, $[\gamma_i]$ for $[I]$ and $[l'_i]$, $[\gamma'_i]$ for $[I']$ in Table 4.2. The computation of inclusion of boundary curves will either result in $[0, 1]$ or *false*. The inclusion result can be *true* only if the boundary curves overlap each other.

Algorithm 1 Compute $[t_{l'}]([I])$

1: **function** COMPUTE $[t_{l'}]([I])([I], [I'])$
2:     **if** $[t_{\gamma'_i}]([\gamma_j])] = [0, 1]$ or $[t_{l'_i}]([\gamma_j])] = [0, 1]$ or $[t_{l'_i}]([l_j])] = [0, 1]$ for each $i = 1, 2$ and each $j = 1, 2$ **then**
3:         $[t_{l'}]([I]) = [0, 1];$
4:     **else**
5:         $[t_{l'}]([I]) = 0;$
6:     **end if**
7: **end function**

This section presented test functions to check the inclusion of one interval in another. Next, we need to find the interval solution for inclusion of two intervals. This solution gives the free intervals in which each point for the robot's next position is guaranteed to avoid the collision.

4.4 Free Interval Computation for Collision Avoidance[4]

In this section, the free interval for the decision-making robot is computed based on bisection of intervals. It is interesting to note that the interval operations described in Section 4.2 use addition/subtraction and division by 2. These operations are very conveniently implemented in embedded platforms. Moreover, hardware architecture design using FPGA would also be possible due to ease of digital computing. Let us now see if an iterative algorithm can be developed that uses interval arithmetic and computes the inclusion of two intervals.

Algorithm 2 computes the interval $[J]$ which is the outer approximation of inclusion. The collision-free intervals are computed by finding the complement of this outer approximation interval (for example, interval $[J]$ as shown in Figure 4.5). The inputs to the algorithm are the intervals $[I]$ and $[I']$ and the values N_r and N_θ, where N_r and N_θ are number of iterations for bisection with respect to r and θ, respectively. The number of iterations N_r and N_θ for r and θ, respectively, are

---

4 Taken from Vyas (2017).

selected based on maximum tolerances $(r_2 - r_1)/2^{N_r}$ and $(\theta_2 - \theta_1)/2^{N_\theta}$ in the respective dimension.

For finding $[J]$, inclusion test of $[I']$ in $[I]$ is computed. In particular, if $[t_{I'}]([I]) = 0$, the two intervals do not intersect and result in no collision; therefore, $[J] = [0, 0][0, 0]$. Algorithm 2 outputs $[J]$, which is the intersection of $[I]$ and $[I']$, the portion which needs to be avoided for safe maneuvers. The interval $[I]$ is bisected with respect to r and then with respect to θ. *It is worth noting that the interval solution avoids those solutions for which intermediate positions at any time between 0 to ΔT may result in collision.* The value of ΔT can be varied according to the operating space so that collision-free intervals can be obtained after finding the inclusion of the intervals. The algorithm runs the inclusion test by computing $[t_{I'}]([I])$. When a collision is detected, i.e. $([t_{I'}]([I]) = [0, 1])$, the interval $[I]$ is bisected in $[L]$ (left interval) and $[R]$ (right interval). Both the intervals are two dimensional given by (4.8), we have $[L] = [L_1] \times [L_2]$ and $[R] = [R_1] \times [R_2]$. Based on the proposed bisection, $[L]$ and $[R]$ are described as follows:

$$[L_i] = [I_i, (I_i + \overline{I_i})/2]$$

$$[R_i] = [(I_i + \overline{I_i})/2, \overline{I_i}]$$

$$[L_{i\%2+1}] = [R_{i\%2+1}] = [I_{i\%2+1}]$$

where $i = 1$ for bisection with respect to r and $i = 2$ for bisection with respect to θ in Algorithm 2. Similarly, $[L]$ is bisected in $[LL] = [LL_1] \times [LL_2]$ and $[LR] = [LR_1] \times [LR_2]$ such that $[LL_i] = [L_i, (L_i + \overline{L_i})/2]$, $[LR_i] = [(L_i + \overline{L_i})/2, \overline{L_i}]$, and $[LL_{i\%2+1}] = [LR_{i\%2+1}] = [L_{i\%2+1}]$ and $[R]$ is bisected in $[RL] = [RL_1] \times [RL_2]$ and $[RR] = [RR_1] \times [RR_2]$ such that $[RL_i] = [R_i, (R_i + \overline{R_i})/2]$, $[RR_i] = [(R_i + \overline{R_i})/2, \overline{R_i}]$, and $[RL_{i\%2+1}] = [RR_{i\%2+1}] = [R_{i\%2+1}]$ for each $i = 1, 2$.

After N_r iterations, the lower limit of $[L_1]$ and upper limit of $[R_1]$ describe $[J_1]$. Similarly, after N_θ iterations, the lower limit of $[L_2]$ and upper limit of $[R_2]$ describe the $[J_2]$. Now, we have $[J] = [J_1] \times [J_2]$. If there is no solution for guaranteed collision avoidance, the output of the algorithm is $[J] = [I]$.

Notice that computing the collision detection and collision-free intervals is possible using only addition and shift operations, thus consuming less power than multiplication and division operations. CORDIC algorithm is helpful in this case. Origin transformation, translation, and rotation can be computed easily using CORDIC as described in Section 4.1 with the help of only adders and shifters. Translation and rotation are used extensively in Algorithm 2 in computation of $[t_{I'}]([I])$ which is the inclusion test function of $[I]$ in $[I']$. In continuation to developing algorithm that requires only digital computations for collision detection and avoidance, Let's understand the interval-based computation of collision detection and free using an illustrative explanation.

Algorithm 2 Collision avoidance

1: **function** COLLISION_AVOIDANCE($[I]$, $[I']$,N_r,N_θ)
2:     Compute $[t_{I'}]([I])$ using 1
3:     **if** $[t_{I'}]([I]) = 0$ **then**
4:         $[J] = [0,0][0,0]$;
5:     **else**
6:         $N \leftarrow N_r$; $n \leftarrow 0$
7:         **for** $i = 1$ to 2 **do**
8:             bisect $[I]$ in $[L]$ and $[R]$
9:             **while** $n < N$ **do**
10:                 bisect $[L]$ in $[LL]$ and $[LR]$ and $[R]$ in $[RL]$ and $[RR]$
11:                 **if** $[t_{I'}]([R]) = 0$ **then**
12:                     $[L] \leftarrow [LL]$; $[R] \leftarrow [LR]$
13:                 **else if** $[t_{I'}]([L]) = 0$ **then**
14:                     $[L] \leftarrow [RL]$; $[R] \leftarrow [RR]$
15:                 **else**
16:                     **if** $[t_{I'}]([LL]) \neq 0$ **then**
17:                         $[L] \leftarrow [LL]$
18:                     **else**
19:                         $[L] \leftarrow [LR]$
20:                   **end if**
21:                   **if** $[t_{I'}]([RR]) \neq 0$ **then**
22:                       $[R] \leftarrow [RR]$
23:                   **else**
24:                       $[R] \leftarrow [RL]$
25:                   **end if**
26:                 **end if**
27:                 $n \leftarrow n + 1$
28:             **end while**
29:             $[J_i] = [L_i, \overline{R_i}]$; $N \leftarrow N_\theta$; $n \leftarrow 0$
30:         **end for**
31:     **end if**
32: **end function**

4.4.1 Illustration for Detecting Collision and Computing Free interval

Figure 4.5 illustrates the two intervals $[I]$ and $[I']$ of a robot and dynamic obstacle, respectively. The inclusion of individual boundary curves is computed such as arc–arc inclusion, arc-line-segment inclusion, and line-segment - line-segment inclusion, to obtain $[t_{I'}]([I])$ using Algorithm 1. Computations of these interval inclusions for digital implementation can be based on lemmas presented in Vyas et al. (2020). In these lemmas, the computation of inclusion of individual boundary curves is performed efficiently using CORDIC-based techniques.

Collision-free interval is computed using branch and bound techniques based on interval arithmetic as explained in Algorithm 2. The 2D mobile robot is represented in polar coordinates form. Thus, Algorithm 2 is implemented by bisection of intervals in iterations with respect to both r and θ. Figures 4.6 and 4.7 show the iterative outcome of bisection of these intervals in iterative manner. At most two intervals in each direction (r and θ) are maintained that represent the inclusion boundary. In each iteration, these intervals shrink and by iteration 6 in illustrative case, the threshold limits of interval dimensions reached. Now, the interval corresponding to collision with respect to each direction r and θ is obtained using infimum and supremum of intervals obtained in iteration 6 as shown in Figure 4.8. The intersection of the colliding intervals in the directions of r and θ renders colliding interval [J] with respect to the origin O (decision-making robot). The complement of interval [J] gives final collision-free interval as shown

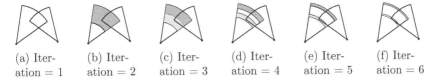

(a) Iteration = 1 (b) Iteration = 2 (c) Iteration = 3 (d) Iteration = 4 (e) Iteration = 5 (f) Iteration = 6

Figure 4.6 An illustration of iterations of bisection with respect to r

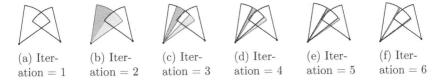

(a) Iteration = 1 (b) Iteration = 2 (c) Iteration = 3 (d) Iteration = 4 (e) Iteration = 5 (f) Iteration = 6

Figure 4.7 An illustration of iterations of bisection with respect to θ

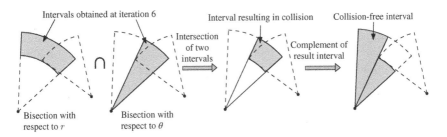

Intervals obtained at iteration 6   Interval resulting in collision   Collision-free interval

Intersection of two intervals   Complement of result interval

Bisection with respect to r Bisection with respect to θ

Figure 4.8 Collision-free interval

in Figure 4.8. In FPGA, the hardware-efficient architecture for Algorithm 2 can be implemented easily since it requires only shift registers for bisecting the intervals and the bisections with respect to r and θ can be implemented in parallel.

4.5 Notes for Further Reading

There exist numerous applications of the CORDIC in varied fields (Meher et al., 2009). Interested readers may refer to the applications of CORDIC in signal processing (Kulshreshtha and Dhar, 2018), control applications (Wills et al., 2012), (Ruppert et al., 2016), robotics (Vyas et al., 2016), and communications (Yao et al., 2018). An architecture developed for the use in WSN (Wireless Sensor Network) (Biswas and Maharatna, 2015) shows that the architecture built over CORDIC has less full adder count as compared to the one developed using other technique.

Many variations of the conventional CORDIC (Volder, 1959) exist and readers may refer to works by Aggarwal et al. (2016), Garrido et al. (2016), and Vachhani et al. (2009) for improvising the requirements such as architecture for reducing semiconductor area, Jaime et al. (2010) and Shukla and Ray (2014) for memory-less technique, scaling-free technique, algorithm for reducing number of iterations. In a recent approach (Meher and Park, 2019), the pipelined architecture for given delay has been developed by combining the recursive and non-recursive forms. Variations based on use of CORDIC in a specific application also exist such as authors in Mahdavi and Timarchi (2019) address speed of CORDIC for computing FFT, while authors Meyer-Bäse et al. (2003) developed CORDIC-based parallel technique for computing Gaussian function. Another interesting work by Luo et al. (2019) addresses computation of exponential and logarithm with arbitrary base using CORDIC which modifies the memory entries pre-computed for the given base. The Verilog code for implementing CORDIC is provided; however, we recommend to readers new to HDL to first study material in Chapter 6 which forms the basis for Verilog coding.

Interval analysis by Moore et al. (2009) is mainly used for global optimization and constraints satisfaction problems. However, interval analysis has also been applied in robotic tasks as seen in Merlet (2009, 2010) and Jaulin et al. (2018). Various robotic tasks such as path planning in Gasparetto et al. (2012), Chauhan et al. (2018), and Jaulin et al. (2001), state estimation in Walter et al. (2001) and Jaulin (2009), collision detection and avoidance in Vyas et al. (2019, 2020), and stability analysis in Jaulin and Le Bars (2012) have been deployed using interval analysis. This chapter presented hardware-efficient implementations of robotic tasks using CORDIC and interval arithmetic. However, there are other research works that deploy hardware-efficient algorithms for robotic tasks. Interested readers can refer to Wan et al. (2021), García et al. (2014), Plancher et al. (2021), and Murray et al. (2016) for more hardware-efficient implementations in robotics.

5

Top-Down Method

In many specific scenarios, the design of the control law is followed by the selection of an embedded platform. It may be the scenario that the processor is already interfacing the peripherals (memory, sensors, and actuators) and the control design is taken at a later stage in the entire system design. The objective is now the exact replication of computations in digital required for controller implementation. Unlike the approach in Chapter 4, the controller design is independent of the selection of embedded platform. Therefore, considering the limitations of embedded implementations, the objective for designing control law is to minimize the memory and computations, or perhaps limit the computations that are free from errors due to digital computations.

One source of error in digital computations is overflow or underflow due to limited word length. Selecting an appropriate word length avoids overflow or underflow errors. This word length selection is particularly important when designing proportional integral derivative (PID) controllers. The integral part of the control law is implemented using an accumulator and can cause overflow or underflow very quickly. In such scenarios, actuator saturation limits aids in selecting the word length for an accumulator design. However, logical computations, like comparison, are another option for designing the control law. One example of digital computations that do not depend on the word length is the signum function. In this chapter, we relate the use of signum function with the sliding-mode controller theory.

This top-down approach requires special attention for control law design and associated analysis. While this chapter evolves sliding-mode concepts for embedded robotic applications, it is applicable for other controller design methods. Similar approaches are applicable for observer designs, which would estimate state and apply a simple feedback control methodology. In summary, the controller or observer design must use computations simple to implement in the digital domain, avoiding substantial processing time and large word lengths, but adequate quantization error that does not affect the stability analysis. This

Embedded Control for Mobile Robotic Applications, First Edition.
Leena Vachhani, Pranjal Vyas, and Arunkumar G. K.
© 2022 The Institute of Electrical and Electronics Engineers, Inc. Published 2022 by John Wiley & Sons, Inc.
Companion website: www.wiley.com/go/vachhani/embeddedcontrolforroboticapp

chapter covers a few controller techniques that result in minimal computations suitable for embedded implementation.

5.1 Robust Controller Design

A general control system is described in Section 1.1. Controller design generally involves measurement of system states and modifying the system inputs based on the measurements to yield desired output. The extent to which the system states are measurable and the input can modify the system behavior is captured by two system properties: observability and controllability of the system. Observability is the system property which indicates whether the current sensor configuration of the system is capable of measuring all the system states. Similarly controllability decides the capability of existing actuators to modify system states from one configuration to other. Depending on these, the controller design is done using various methods. It may be stabilizing the system states to a particular value or tracking a trajectory in the state space. Also, in practical systems the controller may need to tackle the modeling errors or other disturbances affecting the system to yield the desired result. One may refer to Wonham (1974) for a detailed treatment of the subject. In this section, we will introduce a few basic definitions, followed by concepts on state feedback and sliding-mode controllers.

5.1.1 Basic Definitions

Consider the system described by

$$\dot{x} = f(x), \quad x(0) = x_0 \tag{5.1}$$

where $x \in \mathbb{R}^n$ and $f : \mathbb{R}^n \to \mathbb{R}^n$. Here, the assumption is that a unique solution to system given in (5.1) for the given initial condition exists. Also, the systems considered here are assumed to be time-invariant for simplicity. A point $x_e \in \mathbb{R}^n$ is the equilibrium point of the system if it satisfies $f(x_e) = 0$. Without loss of generality the equilibrium point can be assumed to be $x_e = 0$ by shifting the origin of the system as follows. Let $\tilde{x}_e \neq 0$ be the equilibrium point for the system given in (5.1). On applying the change of variable, $y = x - \tilde{x}_e$ and finding the derivative of y gives

$$\dot{y} = \dot{x} = f(x) \tag{5.2}$$
$$= f(y + \tilde{x}_e)$$
$$:= g(y)$$

This gives $g(0) = 0$ and the equilibrium point of the new system, $\dot{y} = g(y)$ is at origin. Further, all the definitions assume that the equilibrium point of the system is at origin unless otherwise stated explicitly. Now, we define the notion of stability of an equilibrium point in the sense of Lyapunov.

Definition 5.1 *(Stability in the sense of Lyapunov)* An equilibrium point $x_e = 0$ of the system (5.1) is said to be stable (in the sense of Lyapunov) if for any $\epsilon > 0$, there exist a $\delta(\epsilon) > 0$, such that

$$\|x(0)\| < \delta \;\Rightarrow\; \|x(t)\| < \epsilon, \quad \forall t > 0 \tag{5.3}$$

This intuitively refers to the fact that when the system is stable, the system states starting at initial conditions sufficiently close to the equilibrium point will remain near the point. Further, the notion of stability is refined for the system where the solution of the system converges to the equilibrium point as time tends to infinity. The definition is as follows.

Definition 5.2 *(Asymptotic stability)* An equilibrium point $x_e = 0$ of the system (5.1) is said to be asymptotically stable if x_e is stable as given in Definition 5.1 and

$$\|x(0)\| < \delta \;\Rightarrow\; \lim_{t \to \infty} x(t) = 0 \tag{5.4}$$

Now, given the stability definitions we need a tool to analyze the stability of the given system. A conventional approach is to integrate the differential equation to obtain a general solution to the system and prove that the conditions for stability definitions are satisfied. But it turns out to be cumbersome and sometimes impossible to obtain an analytical solution. Lyapunov stability theorem defines a method to determine the stability of the given system without obtaining the solution directly. It uses "energy"-like functions defined on the domain of the system to analyze the trajectory of the solutions and defines the conditions for stability. Following theorem states the Lyapunov method for analyzing the stability for a general system.

Theorem 5.1 *(Lyapunov stability theorem)* *Let $x_e = 0$ be an equilibrium point for the system given in (5.1) and $V(x)$ be a continuously differentiable function such that*

$$V(x) > 0 \;\forall x \quad except \; x = 0 \tag{5.5}$$

$$V(0) = 0 \tag{5.6}$$

If $\dot{V}(x) \leq 0$, $\forall x$, then the equilibrium point, $x_e = 0$ is stable.
    If $\dot{V}(x) < 0$, $\forall x$ except $x = 0$, then the equilibrium point, $x_e = 0$ is asymptotically stable.

It is worth noting that inability to find a Lyapunov candidate function doesn't imply that the system is unstable. It only determines whether the system is stable when we are able to define a Lyapunov function satisfying the given conditions.

For general linear systems, the equation given in (5.1) is modified as follows.

$$\dot{x} = Ax, \quad x \in \mathbb{R}^n, \ A \in \mathbb{R}^{n \times n} \tag{5.7}$$

The stability conditions for linear systems are defined with respect to eigen values of the state transformation matrix, A as given in the following definition.

Definition 5.3 **(*Stability of linear systems*)** The system given in (5.7) is asymptotically stable if the real part of all the eigen values(λ_i) of the state transformation matrix, A is negative, $\Re(\lambda_i(A)) < 0$.

Now we introduce state feedback control for systems and analyze the stability using the above definitions.

5.1.2 State Feedback Control

Consider the linear system given in (5.7). The equation is modified as follows to show the most generic form which also depicts the relationship between input to and output from the system.

$$\dot{x} = Ax + Bu \tag{5.8}$$

Here $B \in \mathbb{R}^{n \times p}$ is the input matrix and $u \in \mathbb{R}^{p \times 1}$ is the input. In state feedback, the input u is assigned $-Kx$, where $K \in \mathbb{R}^{m \times n}$ is the gain matrix and x is the state vector. Now, the system dynamics gets modified as follows.

$$\dot{x} = (A - BK)x \tag{5.9}$$

Here the idea is to design the gain matrix, K such that the eigen values of the resultant state transformation matrix, $(A - BK)$ have negative real parts. The basic requirement for arbitrary placement of the eigen values is that the system should be controllable. There exist different methods to design the feedback gain. One way is to calculate the characteristic equation for the modified system, $(A - BK)$ and compare the coefficients of the system to the desired one to obtain the values of the gain. Let us consider the system matrices given below,

$$A = \begin{bmatrix} 0 & 1 \\ -1 & 2 \end{bmatrix} \quad B = \begin{bmatrix} 0 \\ 1 \end{bmatrix}$$

The eigen values of the matrix A are $(1,1)$ and the system is unstable without any input. Now, the state feedback controller is added to the system, that is $u = Kx$ where $K = [k_1 \ k_2]$. The characteristic equation of the modified system $A - BK$ is calculated as follows.

$$
|sI - (A - BK)| = \begin{vmatrix} s & -1 \\ 1 - k_1 & s - 2 - k_2 \end{vmatrix}
$$
$$
= s^2 - s(2 + k_2) + (1 + k_1) \tag{5.10}
$$

Let the eigen values of the desired system be $(-1, -2)$, which is negative and render the system stable. The characteristic equation of the desired system is calculated to be $s^2 + 3s + 2$. On comparing the coefficients of (5.10) with the desired characteristic equation, the gain matrix is calculated to be $K = [1\ 5]$. This method is easy for a smaller system, but the calculation gets tedious once the dimension increases. Other methods perform a linear transformation of the system to get the controller canonical form of the system and design the controller gains for the transformed system easily. Then, these gains can be converted back to the original system. Another way is to use Ackermann's formula for calculating the gain. One may refer to Williams and Lawrence (2007) for further reading.

A single integrator system is one of the simplest models for a robotic platform where the input is velocity and the position is modified according to the integral of the input. Also, a relation between unicycle model and single integrator system is explained in Section 1.2.1. An orientation-independent robotic platform in 2D with two inputs, one for controlling the position in x direction and another for controlling the movement in y direction, can also easily be modeled as unit with two single integrator systems. Thus, the analysis of a simple single integrator gives an insight into the controller performance. Let us look at the application of state feedback system on a single integrator system. Consider the single integrator system with states $x(t)$ and input $u(t)$ given below.

$$\dot{x}(t) = u(t) \tag{5.11}$$

Now let us use the state feedback to design the input with gain matrix K as follows:

$$u(t) = -Kx(t) \tag{5.12}$$

Let us analyze the system stability using Lyapunov stability analysis. Consider the candidate Lyapunov function to be

$$V(t) = \frac{1}{2}x^2(t) \tag{5.13}$$

The candidate Lyapunov function is positive for all values of $x(t)$ and is zero only when $x(t)$ is zero. On differentiating the candidate Lyapunov function, the derivative turns out to be

$$\dot{V}(t) = x(t)\,\dot{x}(t)$$
$$= x(t)\,u(t)$$

On substituting the state feedback input (5.12), the derivative of the candidate Lyapunov function is

$$\dot{V}(t) = -Kx^2(t)$$

The derivative of the Lyapunov function is negative definite for positive values of K, or $K > 0$. The system is stable for the equilibrium point $x = 0$. Hence, state

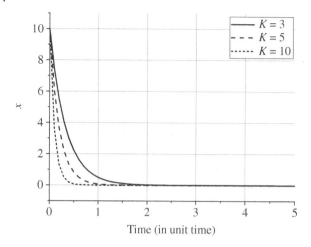

Figure 5.1 Stabilization of a single integrator using state feedback

feedback control is very simple but effective method for stabilizing the system. A simulation for stabilization of single integrator system using state feedback is shown in Figure 5.1. The initial value of the state, $x(0) = 10$ is considered for the simulation. The stabilization of the system for three different values of the gain, $K = 3, 5, 10$ is considered. The value of the state converges to origin smoothly using state feedback control. Also, it can be observed that as the value of the gain increases, the rate of convergence increases. Now, consider the scenario of the same system with disturbance affecting the system through the input channel. The system dynamics with the disturbance d affecting the system is given by

$$\dot{x}(t) = u(t) + d \tag{5.14}$$

Let us apply the same controller given in (5.12) to this system, (5.14). Now, the system dynamics modifies as follows

$$\dot{x}(t) = -Kx(t) + d \tag{5.15}$$

Here, we may use the same Lyapunov function as given in (5.13), and it can be observed that the stabilization cannot be proved. As conveyed earlier, it is important to note that if a candidate Lyapunov function does not show stability, it doesn't prove that the system is unstable. It only indicates the chosen candidate Lyapunov function cannot prove the stability of the system. Figure 5.2 depicts a simulation of the system given in (5.15). The initial condition of the system state is chosen to be $x(0) = 10$. The disturbance is chosen to be uniform random variable in the interval $[-3, 3]$. The state feedback control is simulated for three different values of the gain, $K = 3, 5, 10$. The figure in the inset shows that the state feedback control is not able to stabilize the system and once the system state reaches near

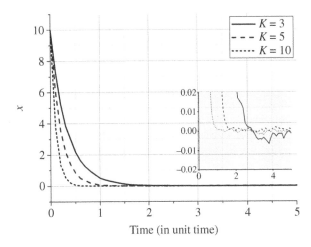

Figure 5.2 Effect of state feedback control on a single integrator with disturbance affecting the input channel

the origin. The disturbance drives away the system state. It should be noted that as the value of gain increases, the effect of disturbance reduces. Thus choosing high values of the gain is an effective method for stabilizing the system affected by disturbance. But the constraints like actuator saturation and power requirements disallow the designer from choosing high gain values in practical scenarios. A robust control technique is introduced in Section 5.1.3 which can address these challenges.

Note that a direct application of linear state feedback theory is not possible on unicycle model that we have discussed. It is because of the fact that the linearized model of the unicycle is not fully controllable, and thus any linear controller is not capable of stabilizing the system, even around the point of linearization. Also, the stabilization of unicycle robots using smooth time-invariant control methods is impossible due to the results shown by Brockett (1983). There are different controllers developed for trajectory tracking of unicycle robots, but need complicated formulations. For a detailed treatment of this, one may refer to De Luca et al. (2001).

5.1.3 Sliding-Mode Control

A major problem while dealing with controller design for mobile robot applications is the inaccuracies of the system modeling due to the complexity of the system, unmodeled dynamics or the effect of variation of the system parameters during the operation. Another practical issue arises when the system performance is affected by different disturbances that arise in the environment or the actuation errors arise within the system itself. A controller design should thus anticipate these issues and design a robust controller which can tackle these problems. Sliding-Mode Control (SMC) is a robust control technique that can be applied to a variety of control applications. The biggest advantage of the SMC is the robustness of the control law toward the disturbances that affect the system through input channel.

SMC is a discontinuous control technique which is used for robust control of the systems. Consider a trivial example of a discontinuous controller,

$$u(t) = -K \, \text{sgn}(x(t)) \tag{5.16}$$

for the system described by (5.11). Here $x(t)$ is the state of the system, $u(t)$ is the input or control applied, and K a constant gain. Now, the objective is to show that the simple control law given by (5.16) has finite-time convergence and robustness properties. Let us analyze the stability of the system around origin using Lyapunov method. For the system described by (5.11), let the candidate Lyapunov function be

$$V(t) = \frac{1}{2}x^2(t) \tag{5.17}$$

The candidate Lyapunov function is positive for all values of $x(t)$ and is zero only when $x(t)$ is zero. On differentiating (5.17) and substituting the dynamics of the system in (5.11), we get

$$\dot{V}(t) = x(t)\dot{x}(t)$$
$$= x(t)u(t) \tag{5.18}$$

Substituting the input, $u(t)$ from (5.16) and further reduction yields

$$\dot{V}(t) = -Kx(t) \, \text{sgn}(x(t))$$
$$= -K|x(t)|$$
$$= -K\sqrt{(2V(t))} \tag{5.19}$$

Here, $\dot{V}(t) < 0$ for $K > 0$, and hence the discontinuous input is capable of driving the system to origin. Now, let us calculate the time for convergence from (5.19). On rearranging and integrating the derivative we get,

$$\int_{V(0)}^{0} \frac{1}{\sqrt{V(t)}} dv = -\sqrt{2}K \int_{0}^{T} dt \tag{5.20}$$

Here $V(0)$ is the initial value of the Lyapunov function that depends on the initial conditions of the system. Also, T is the time at which the Lyapunov function converges to zero. On further reduction the integral in (5.20) reduces as follows

$$\left[2\sqrt{V(t)}\right]_{V(0)}^{0} = -\sqrt{2}K\,[T]_{0}^{T} \tag{5.21}$$

Reducing the equation further, the time for convergence for the Lyapunov function or the system is given by

$$T = \frac{\sqrt{2V(0)}}{K} \tag{5.22}$$

This shows that the system converges to zero within finite time, T as given in equation (5.22). This is a stronger result than the general asymptotic stability that we have seen. The result is demonstrated in Figure 5.3 for a system with initial conditions $x(0) = 10$ and for different values of K. The simulation is performed in MATLAB© for $K = 1, 2, 5$, and 10. It can be observed that as the value of K increases, the faster the convergence which is in accordance with (5.22).

Now, let us analyze the robustness of the system with discontinuous input against a bounded disturbance affecting the system through input channel. Consider the system in the example given below.

$$\dot{x}(t) = u(t) + d(t) \tag{5.23}$$

Here, $d(t)$ is the bounded disturbance affecting the system through input channel. Let the bound on the disturbance be $|d(t)| < D$. The stability of the system is analyzed by choosing the same candidate Lyapunov function given in (5.17). On differentiating the Lyapunov function we get

$$\dot{V}(t) = x(t)\dot{x}(t)$$
$$= x(t)\,(u(t) + d(t)) \tag{5.24}$$

Substituting the discontinuous controller, (5.16) used in the previous example

$$\dot{V}(t) = x(t)\left(-K\,\mathrm{sgn}(x(t)) + d(t)\right) \tag{5.25}$$

Figure 5.3 Stabilization of a single integrator using discontinuous input

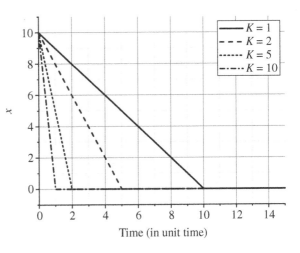

Substituting the bound for disturbance in (5.25) yields,

$$\dot{V}(t) \leq -x(t)\,\text{sgn}(x(t))\,(K - D)$$
$$= |x(t)|\,(K - D)$$
$$= (K - D)\,\sqrt{(2V(t))} \tag{5.26}$$

From (5.26), the system is finite time stable when $(K - D) > 0$, or $K > D$. A similar analysis given in (5.20), can be performed to calculate an upper bound on the time for convergence for the system states to origin. The simulation results for a single integrator system affected by a matched disturbance modeled as uniform random variable is shown in Figure 5.4. The bound for the matched disturbance, $D = 2$. The results for $K = 3, 5$, and 10 are plotted. It can be observed that the system converges to origin irrespective of the matched disturbance.

The discontinuous control law hence proves to be an efficient method for driving the system given in (5.23) to origin. But, on further inspection we can see that the system considered in the example is fully actuated or the number of states and that of control inputs are same. For a general mobile robot configuration, this may not be valid. The number of states of the system will be larger than the number of control inputs, thus under-actuated. Hence, this control law is not directly applicable in those cases. The control of under-actuated system needs an introduction to the concept of sliding surface. An intuitive discussion on the idea of the sliding surface and the effect of discontinuous input is made further to motivate the SMC law design.

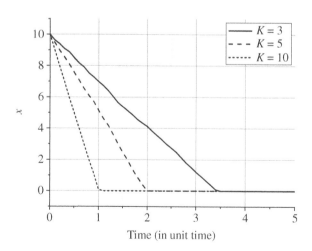

Figure 5.4 Stabilization of a single integrator affected by disturbance ($D = 2$), using discontinuous input

In the previous example, the stabilization of a single-state system with a discontinuous input is demonstrated. The proposed input is able to drive the system to origin within finite time, even in the presence of matched disturbance. Now, consider a system with n-dimensional state vector, which is controllable. A single input of similar nature can drive the system dynamics to a surface which is the function of system states within finite time. Also, the control input will restrict the system dynamics to the proposed surface and reduce the effective dimension of the system to $(n - 1)$. The reduced order system or the sliding surface is designed in such a way that once the system trajectory is restricted to this space, the system stabilizes to equilibrium asymptotically (or may be within finite time). This idea is further demonstrated using a simple under-actuated double integrator system discussed in (1.3). The system consists of two states and an input, as given below.

$$\dot{x}_1(t) = x_2(t)$$
$$\dot{x}_2(t) = u(t) \tag{5.27}$$

The state space representation of the double integrator described by (5.27) is given by

$$\dot{x} = Ax + Bu \tag{5.28}$$

where $x \in \mathbb{R}^2, A \in \mathbb{R}^{2 \times 2}, B \in \mathbb{R}^{2 \times 1}$, and u is a scalar where,

$$x = \begin{bmatrix} x_1 \\ x_2 \end{bmatrix} A = \begin{bmatrix} 0 & 1 \\ 0 & 0 \end{bmatrix}$$
$$B = \begin{bmatrix} 0 \\ 1 \end{bmatrix} \tag{5.29}$$

The variable t is dropped for brevity. The double integrator system is an under-actuated system with two states and one input. Let us investigate if the system is controllable. The controllability matrix of the system is given by

$$C = \begin{bmatrix} 0 & 1 \\ 1 & 0 \end{bmatrix} \tag{5.30}$$

The rank of the controllability matrix C is 2, which is full rank and hence the system is fully controllable. Now, let us design a sliding-mode controller to stabilize the system. As explained in the previous discussion, a sliding surface is to be designed, such that once the system dynamics is restricted to the sliding surface, the system stabilizes asymptotically. Let the sliding surface be

$$S(x) = \lambda x_1 + x_2 \tag{5.31}$$

where λ is a design parameter. Now we need to prove the convergence of the system dynamics to the proposed sliding surface within finite time. We again resort to Lyapunov analysis to understand the dynamics and consider a candidate Lyapunov function as follows:

$$V = \frac{1}{2}S^2$$
$$= \frac{1}{2}\left(\lambda x_1 + x_2\right)^2 \tag{5.32}$$

On differentiating the equation in (5.32), the dynamics of the Lyapunov function renders

$$\dot{V} = \left(\lambda x_1 + x_2\right)\left(\lambda \dot{x}_1 + \dot{x}_2\right) \tag{5.33}$$

From (5.27) and (5.33), we get

$$\dot{V} = \left(\lambda x_1 + x_2\right)\left(\lambda x_2 + u\right) \tag{5.34}$$

Let the sliding-mode controller to drive the system dynamics to the sliding surface be

$$u = -\lambda x_2 - K\text{sgn}(S(x)) \tag{5.35}$$

Substituting the control input from (5.35) in (5.34) and further reduction renders

$$\dot{V} = -K\,\text{sgn}(S(x))\left(\lambda x_1 + x_2\right)$$
$$= -K|\lambda x_1 + x_2|$$
$$= -K\sqrt{2V} \tag{5.36}$$

Equation (5.36) ensures that the system dynamics converges to the sliding surface within finite time. The finite time calculations are very similar to the examples provided above. Now, we need to analyze the system dynamics on the sliding surface. When $(S(x) = 0)$, we can replace $x_2 = -\lambda x_1$.

$$\dot{x}_1 = -\lambda x_1 \tag{5.37}$$

For the positive values of λ, the state x_1 converges to zero asymptotically (finite time convergence can also be shown). Also, the system dynamics restricted on the sliding surface ensures that $x_2 = -\lambda x_1 = 0$.

Thus, the system dynamics under the influence of an SMC law is divided into two phases: the reaching phase and sliding phase. Reaching phase denotes the duration when the system dynamics is out of the sliding surface and the control law is driving the system toward the sliding surface. The sliding phase refers to the time period after the system dynamics reaches the sliding surface and is restricted on to the sliding surface. Also, note that the reaching phase has to be performed within finite time as it is just an intermediary phase to bring the system trajectories to the sliding surface where the desired system states or the entire system itself stabilizes asymptotically. During the sliding phase, the dynamics of the system is driven to origin by the reduced order dynamics. The design of the sliding

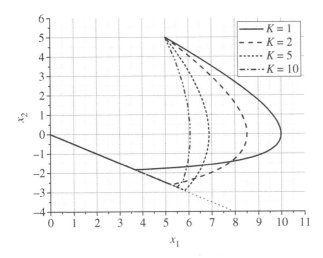

Figure 5.5 Stabilization of a double integrator by a sliding-mode controller

surface is such that restriction of system trajectories on to the surface ensures the stabilization of system or desired state variables.

Figure 5.5 shows the phase plane for a set of simulations of the proposed SMC law applied on a double integrator with initial conditions $x_1(0) = 5$ and $x_2(0) = 5$. The simulation is performed for different values of K as shown in the figure. The value of λ is chosen to be 0.5. The curved part of the phase plane trajectories denotes the reaching phase to the sliding surface. The sliding surface is a line, marked with dots from $(0, 0)$ to $(8, -4)$ in the figure. It can be observed that the system trajectories converge and remain on the sliding surface from the figure. The sliding phase of all the simulations is overlapping the sliding surface plot in the figure. For more clarity, the dynamics of the individual states of the double integrator is plotted in Figure 5.6 for $K = 1$; notice the stabilization of individual states to origin.

Now, just like the example of the single integrator given above, the sliding-mode controller is robust against any matched disturbance affecting the input channel. A similar analysis for single integrator can be performed for calculating the value of gain required to suppress the effect of disturbance.

5.1.4 Sliding Surface Design for Position Stabilization in 2D

In this section, we will discuss a sliding-mode controller design for position stabilization of the 2D unicycle robot, discussed in Section 1.2.1. Navigation query typically stated for source and destination positions. Hence, the control objective

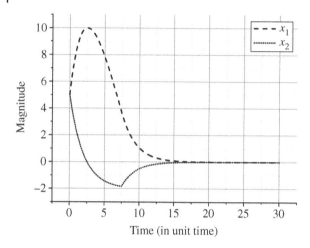

Figure 5.6 Double integrator states trajectory under the influence of sliding-mode controller

is to stabilize position only as against entire state stabilization which includes position as well as orientation. Moreover, it is expensive for the embedded implementations to rely on entire state feedback and is challenging to implement observer designs due to computation overheads. Thus, a controller design that relies on limited feedback and needs minimal computations is ideal for such a scenario. A sliding-mode controller that relies on coarse bearing feedback, and is robust to actuation errors and disturbances serve ideal for embedded implementation on inexpensive platforms. The aim for developing a control law is to bring the mobile robot arbitrarily near to the goal position, irrespective of the initial pose of the robot using bearing feedback.

The kinematics of a unicycle robot moving in 2-D plane is discussed in (1.6). Figure 5.7 shows a unicycle robot in a 2D plane with reference to an inertial frame,

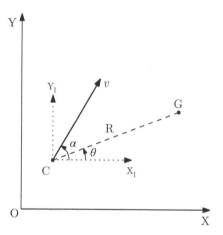

Figure 5.7 Engagement geometry between current position of the robot and the target position

XY with origin **O**. The current position of the robot is marked in the figure as **C**, with pose (x, y, α) with reference to the kinematics given in (1.6). In the figure, the heading of the robot is shown with a vector v, where it also denotes the linear velocity of the robot. A translated inertial frame, $X_I Y_I$ is shown at **C**. The orientation of the robot is α, the angle between X_I and the velocity vector. The target position is denoted with point **G** in the figure. The stabilization of the robot in 2D requires the controller to reduce the radial distance between the robot and the goal position. Also, the objective of the controller design is to rely on coarse measurements of relative bearing feedback, so that it is easy to implement on a practical system where the methodology to compute bearing is available using sensor measurements. Hence, it is more meaningful to use polar coordinate system, (R, θ), to represent the relative distance and bearing angle between the goal position and the robot location. Figure 5.7 depicts these variables where $R(t) \in [0, \infty)$ denotes the radial distance from the current position of the robot to the goal point and $\theta(t) \in (-\pi, \pi)$ is the azimuthal angle to the goal. The evolution of the trajectory of the robot with respect to the origin under the influence of velocity inputs, (v, ω), is represented using dynamics of the variables (R, θ, α). The dynamics can be derived to be as given in equation (5.38).

$$\dot{R} = -v\cos(\alpha - \theta)$$
$$R\dot{\theta} = -v\sin(\alpha - \theta) \tag{5.38}$$
$$\dot{\alpha} = \omega$$

Now, let us design a sliding-mode controller to stabilize the robot position to goal point. Note that the aim is to design a sliding surface, to which the restriction of robot kinematics results in reduction of R. Consider the sliding surface given below.

$$S = \{(R, \theta, \alpha) : \sigma = 0\} \quad \text{where,} \ \sigma \ = \ \alpha - \theta \tag{5.39}$$

Let us analyze the geometry of the sliding surface. For any instantaneous robot position, $\sigma = 0$ denotes the velocity vector aligned with the line-of-sight vector $\overrightarrow{CG}$, between the robot and the goal position. By forcing the velocity vector on the line of sight orienting the robot direction toward the target point reduces the radial distance R. Now, we will prove the stabilization with the help of Lyapunov analysis. Let the candidate Lyapunov function V be described by

$$V = \frac{1}{2}\sigma^2$$
$$= \frac{1}{2}(\alpha - \theta)^2 \tag{5.40}$$

The proposed function, V is positive for all values of σ and reduces to zero when $\sigma = 0$. The derivative of V is given by,

$$\dot{V} = \sigma\dot{\sigma}$$
$$= \sigma\left(\dot{\alpha} - \dot{\theta}\right)$$
$$= \sigma\left(\omega + \frac{v}{R}\sin(\alpha - \theta)\right)$$
$$= \sigma\left(\omega + \frac{v}{R}\sin(\sigma)\right) \tag{5.41}$$

Now, we choose the control law as given by (5.42)

$$\omega = -K\,\mathrm{sgn}(\alpha - \theta) \tag{5.42}$$

From (5.42) and (5.41), we get

$$\dot{V} = \sigma\left(-K\,\mathrm{sgn}(\alpha - \theta) + \frac{v}{R}\sin(\sigma)\right)$$
$$= \sigma\left(-K\,\mathrm{sgn}(\sigma) + \frac{v}{R}\sin(\sigma)\right) \tag{5.43}$$

The value of $(\alpha - \theta) \in [-\pi, \pi]$ and hence

$$\mathrm{sgn}(\sigma) = \mathrm{sgn}(\alpha - \theta)$$
$$= \mathrm{sgn}(\sin(\alpha - \theta)) \tag{5.44}$$

Substituting (5.44) in (5.43), we get

$$\dot{V} = -\sigma\,\mathrm{sgn}(\sigma)\left(K - \frac{v}{R}|\sin(\sigma)|\right)$$
$$\leq -|\sigma|\left(K - \frac{v}{R}\right)$$
$$= -\sqrt{2V}\left(K - \frac{v}{R}\right) \tag{5.45}$$

By choosing $K > v/R$, $\dot{V} < 0$ and the robot kinematics reaches the sliding plane $(\sigma = 0)$ in finite time. A good choice is $K = (\kappa v)/R_c$, where R_c is the radius of the circle centered at the goal position and $\kappa > 1$. This will ensure that the derivative of $V(t)$ remains negative for $R > R_c$ from equation (5.45). Further the derivative can be written as follows.

$$\dot{V} \leq -\eta\sqrt{2V} \quad \forall R > R_c \tag{5.46}$$

where, $\quad \eta > (\kappa - 1)\dfrac{v}{R_c} \tag{5.47}$

The finite time convergence to the sliding surface when $R > R_c$ is ensured by (5.47), and the time for convergence can be calculated by the same method used in (5.20). Once the system trajectory is restricted to sliding surface and if $R > R_c$, the dynamics of the radial distance is modified as follows.

$$\dot{R} = -v \tag{5.48}$$

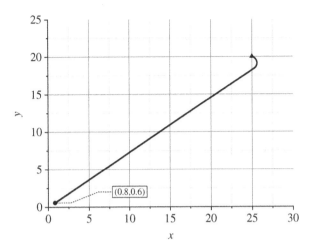

Figure 5.8 Trajectory taken by a unicycle vehicle in 2D using proposed control law

On further analysis of equation (5.48), it is observed that the system position reaches within R_c circle centered at the goal position within finite time. A simulation for the same is depicted in Figure 5.8. The initial pose of the unicycle robot is $(25, 20, -20°)$, marked with a triangle. The robot is assumed to have unit velocity and R_c is chosen to be 1 unit. The design parameter κ is chosen to be 1. The final position of the robot is marked in the figure using a circle and is at $(0.8, 0.6)$, which is inside the unit circle. We extend this control algorithm to perform position stabilization of a vehicle maneuvering in 3D in Section 5.1.5.

5.1.5 Position Stabilization for a Vehicle in 3D

A kinematic model of a generic robot in 3D is given by (1.39). One can look forward to stabilizing all the states; however, it becomes a tougher control problem as the number of control inputs is lesser than the number of state variables. Moreover, non-linearity involved in the model makes the control problem more challenging. One way to deal with the stabilization problem is to stabilize only the position (orientation can have any final value). Further, the control problem can be simplified by understanding the functionality needed from the mobile robot (vehicle). In the defined functionality, certain state variables can be either not interacting or can be decoupled. Along these lines, let us consider a 3D Dubin's vehicle for position stabilization with following assumptions.

1. The robot only has surge velocity, u and the sway and heave velocities, $v = w = 0$.
2. The robot is assumed to be roll stabilized ($\gamma = 0$ and $\dot{\gamma} = 0$ always).

3. The angular velocity inputs are considered in the inertial frame rather than the body frame. Thus the inputs are u_α and u_β which drives the Euler angle rates $\dot{\alpha}$ and $\dot{\beta}$, respectively.

Substituting these conditions in the generic model given in (1.39) yields the kinematic model of the robot as follows.

$$
\begin{aligned}
\dot{x} &= u C_\alpha C_\beta \\
\dot{y} &= u S_\alpha C_\beta \\
\dot{z} &= u S_\beta \\
\dot{\alpha} &= u_\alpha + d_\alpha \\
\dot{\beta} &= u_\beta + d_\beta
\end{aligned}
\tag{5.49}
$$

where the notations $C_* = \cos(*)$ and $S_* = \sin(*)$. Also, here d_α and d_β are the matched disturbances bounded by D_α and D_β, respectively. In particular, $|d_\alpha| < D_\alpha$ and $|d_\beta| < D_\beta$.

Similar to the position stabilization of 2D vehicle in Section 5.1.4, we keep the objective to design a sliding control law to stabilize the position of the robot using minimal sensing and processing. The control law design depends on the coarse relative bearing information toward the target position with respect to the robot, and thus the representation of the target or goal position using spherical coordinates is useful for further analysis. Figure 5.9 depicts a robot in a 3D inertial frame XYZ. The origin of the inertial frame is marked with point **O** and the current robot location is shown with point **C**. The target position is denoted by point **G** and is described relative to the robot location **C** using spherical coordinates, $(R(t), \theta(t), \phi(t))$ at an instant t. A translated inertial coordinate frame $X_I Y_I Z_I$ is shown at the robot position **C** to illustrate these variables. $R(t) \in [0, \infty)$ denotes the radial distance, $\theta(t) \in (-\pi, \pi)$ is the azimuth angle, and $\phi(t) \in (-\pi/2, \pi/2)$ is the elevation angle of the goal point with respect to the robot position, **C**. The heading direction of the robot is marked with a vector u, which is the x-direction of the body frame. Now, the evolution of the relative variables of goal position with respect to the system trajectory (5.49), is given in (5.50).

$$
\begin{aligned}
\dot{R} &= -u C_{(\alpha-\theta)} C_\phi C_\beta - u S_\phi S_\beta \\
\dot{\theta} &= -\frac{1}{R C_\phi} \left[u S_{(\alpha-\theta)} C_\beta \right] \\
\dot{\phi} &= \frac{1}{R} \left[-u C_\phi S_\beta + u C_{(\alpha-\theta)} S_\phi C_\beta \right] \\
\dot{\alpha} &= u_\alpha + d_\alpha \\
\dot{\beta} &= u_\beta + d_\beta
\end{aligned}
\tag{5.50}
$$

To address the objective of position stabilization (reaching to the goal position), the control law aims for the following: the robot is guaranteed to reach sufficiently

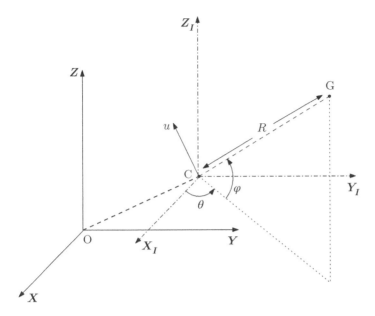

Figure 5.9 3D Dubin's vehicle kinematics in an inertial frame

close to the goal position even in the presence of matched disturbances. Hence a
control law is developed to bring the robot from any radial distance, R, within a
sphere S_R of radius R_c centered at the goal position, where R_c is a design parameter
that signifies a threshold distance close to the goal position. An SMC law for the
system described by (5.50) is developed next, which meets the aforementioned
objectives.

Let us design a sliding surface for the system given in (5.50). The objective of
the design is such that forcing the dynamics of the system onto the suggested sliding surface reduces the distance between the robot and the goal position. Now,
consider the sliding surface given below.

$$S = \{(R, \theta, \phi, \alpha, \beta) : \sigma = 0\} \quad \text{where}$$

$$\sigma = \begin{bmatrix} \phi - \beta \\ \theta - \alpha \end{bmatrix} \tag{5.51}$$

An illustration of the sliding surface is shown in Figure 5.10. Sliding surface
defined by (5.51) is the intersection of two surfaces. The surface, $\phi - \beta = 0$, shown
in light shade, describes an inverted cone at the current robot position passing
through the goal point. The other sliding surface $\theta - \alpha = 0$, shown using dark
shade in Figure 5.10, is the plane perpendicular to the $X_I Y_I$ plane passing through
the robot's current position and the goal point. The intersection of these two

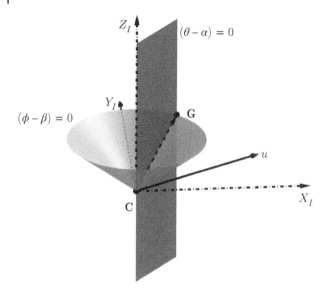

Figure 5.10 Illustration of the sliding surfaces on engagement geometry

surfaces results in the sliding surface S given by (5.51), which is the line-of-sight vector $\overrightarrow{\mathbf{CH}}$. The objective is to bring the system dynamics to this sliding surface ($S = 0$) in finite time. Once the robot aligns to the sliding surface, the resultant dynamics is derived from (5.50) with $\beta = \phi$ and $\alpha = \theta$. Therefore, the trajectory of the resultant system is given by

$$\dot{R} = -u$$
$$\dot{\theta} = \dot{\phi} = 0$$
$$\dot{\alpha} = u_\alpha + d_\alpha$$
$$\dot{\beta} = u_\beta + d_\beta$$

(5.52)

The resultant dynamics evidently shows that the radial distance between the robot and origin reduces ($\dot{R} = -u$) and approaches to zero. In order to derive the control law and show the finite time convergence of system trajectories to the sliding surface, we consider the candidate Lyapunov function as

$$V = \frac{1}{2}\sigma^\mathsf{T}\sigma$$
$$= \frac{1}{2}\left[(\phi - \beta)^2 + (\theta - \alpha)^2\right]$$
$$= V_1 + V_2$$

(5.53)

where

$$V_1 = \frac{1}{2}(\phi - \beta)^2 \quad \text{and} \tag{5.54}$$

$$V_2 = \frac{1}{2}(\theta - \alpha)^2 \tag{5.55}$$

The function V_1 is zero on the sliding surface $(\phi - \beta) = 0$ and positive otherwise. Similarly, the function $V_2 > 0$ for all values of $(\theta - \alpha)$ except on the sliding surface where it remains zero. Thus, the proposed Lyapunov function, $V = 0$ for $\sigma = 0$ and $V > 0$, otherwise. The derivative of V along the trajectory of the system is given by,

$$\dot{V} = \dot{V}_1 + \dot{V}_2, \tag{5.56}$$

where

$$\dot{V}_1 = (\phi - \beta)\left(\dot{\phi} - u_\beta - d_\beta\right) \quad \text{and} \tag{5.57}$$

$$\dot{V}_2 = (\theta - \alpha)\left(\dot{\theta} - u_\alpha - d_\alpha\right) \tag{5.58}$$

Equations (5.50) and (5.57) render

$$
\begin{aligned}
\dot{V}_1 &= (\phi - \beta)\left[\frac{1}{R}\left(uC_{(\alpha-\theta)}S_\phi C_\beta - uC_\phi S_\beta\right) - u_\beta - d_\beta\right] \\
&\leq -|\phi - \beta|\operatorname{sgn}(\phi - \beta)\left[u_\beta - D_\beta - \frac{u}{R}\right] \\
&= -\sqrt{2V_1}\left[u_\beta \operatorname{sgn}(\phi - \beta) - \left(D_\beta + \frac{u}{R}\right)\operatorname{sgn}(\phi - \beta)\right]
\end{aligned}
\tag{5.59}
$$

Let a choice of control input for pitch rate be

$$u_\beta = K_\beta \operatorname{sgn}(\phi - \beta) \tag{5.60}$$

On substituting (5.60) in (5.59),

$$\dot{V}_1 \leq -\sqrt{2V_1}\left[K_\beta - \left(D_\beta + \frac{u}{R}\right)\operatorname{sgn}(\phi - \beta)\right] \tag{5.61}$$

For,

$$K_\beta > \left(D_\beta + \frac{u}{R}\right)\operatorname{sgn}(\phi - \beta) \tag{5.62}$$

$\dot{V}_1$ is negative definite. The inequality (5.62) is satisfied for

$$
\begin{aligned}
K_\beta &> \frac{u}{R} + D_\beta \\
&= \frac{u}{R_c} + D_\beta
\end{aligned}
\tag{5.63}
$$

where R_c is defined as follows.

$$R > R_c > 0 \tag{5.64}$$

Now, from (5.62), let

$$\eta_1 = \min \left(K_\beta - \left(D_\beta + \frac{u}{R} \right) \text{sgn}(\phi - \beta) \right) \tag{5.65}$$

Substituting (5.65) in (5.61),

$$\dot{V}_1 \leq -\eta_1 \sqrt{2V_1} \tag{5.66}$$

Hence the condition for finite time stability as given in Shtessel et al. (2014) is satisfied and V_1 will converge to zero in finite time under the control law, u_β given by (5.60).

Now let us consider the dynamics of V_2 given in (5.58).

$$\dot{V}_2 = (\theta - \alpha) \left(-\frac{1}{RC_\phi} \left[uS_{(\alpha-\theta)}C_\beta \right] - u_\alpha - d_\alpha \right) \tag{5.67}$$

Both $C_\phi > 0$ and $C_\beta > 0 \ \forall \ \phi, \beta \in (-\pi/2, \pi/2)$. Therefore, (5.67) gives

$$\dot{V}_2 = (\theta - \alpha) \left(-K_\phi S_{(\alpha-\theta)} - u_\alpha - d_\alpha \right)$$
$$= (\theta - \alpha) \left(K_\phi S_{(\theta-\alpha)} - u_\alpha - d_\alpha \right) \tag{5.68}$$

where

$$K_\phi = \frac{u}{R} \frac{C_\beta}{C_\phi} > 0 \tag{5.69}$$

From (5.66), we know that $\phi = \beta$ will be achieved in finite time. The value of K_ϕ reduces to u/R once the system trajectory converges to the sliding surface $(\phi - \beta = 0)$. Substituting this condition to (5.68) we get,

$$\dot{V}_2 = |\theta - \alpha| \, \text{sgn}(\theta - \alpha) \left(K_\phi S_{(\theta-\alpha)} - u_\alpha - d_\alpha \right)$$
$$\leq -\sqrt{2V_2} \left(u_\alpha \text{sgn}(\theta - \alpha) - (D_\alpha + \frac{u}{R}) \right) \tag{5.70}$$

Let us consider the yaw rate input to be

$$u_\alpha = K_a \text{sgn}(\theta - \alpha) \tag{5.71}$$

Substituting (5.71) in (5.70),

$$\dot{V}_2 \leq -\sqrt{2V_2} \left(K_a - \left(D_\alpha + \frac{u}{R} \right) \right) \tag{5.72}$$

For $\dot{V}_2$ negative definite, K_a must satisfy

$$K_a > D_\alpha + \frac{u}{R_c} \tag{5.73}$$

where R_c is defined by (5.64). Therefore, choice of u_α given by (5.71) renders $\dot{V}_2$ negative definite. Hence, the finite time convergence of pitch to the conical sliding surface, ($\beta = \phi$) will guarantee the validity of the condition for K_α given in (5.73). Thus again, the choice of u_α given by (5.71) renders $\dot{V}_2$ negative definite. Let

$$\eta_2 = \min \left(K_\alpha - \left(D_\alpha + \frac{u}{R_c} \right) \right) \tag{5.74}$$

Substituting (5.74) in (5.72) yields

$$\dot{V}_2 \leq -\eta_2 \sqrt{2V_2} \tag{5.75}$$

Hence the condition for finite time stability (Shtessel et al., 2014) is satisfied. In other words, V_2 approaches zero in finite time under the control law, u_α given by (5.71). Hence the control inputs chosen as in (5.60) and (5.71) ensure that both $\dot{V}_1$ and $\dot{V}_2$ are negative definite; hence, by (5.56), $\dot{V} < 0$ for all states not in S. The finite time convergence of the robot's orientation to the direction of line of sight vector (vector **GC** in Figure 5.10) follows trivially from the conditions for finite time convergence of orientation to the sliding surfaces ($\phi = \beta$) and ($\alpha = \theta$).

Remark 5.1 The SMC method developed in this chapter only controls the yaw and pitch steering. The linear velocity of the robot is untouched; thus, any linear velocity profile can be used to control the robot. The maximum value of linear velocity profile is used for computing bounds of K_α and K_β.

Once the robot reaches the sliding surface, the reduced order dynamics of the robot as in (5.52) ensures that the robot approaches sphere $\mathcal{S}_\mathcal{R}$. A simulation of the algorithm is depicted in Figure 5.11. The robot is considered to have unit surge velocity and the initial pose of the robot is $(25, 20, 15, 45°, 80°)$. In the figure, a triangular marker denotes the initial position. The disturbance is simulated as a uniform random variable with $D_\alpha = D_\beta = 0.2$. The design parameter R_c was chosen to be unit distance. K_α and K_β are chosen according to the proposed algorithm. It can be seen that the robot converges to unit sphere from the graph.

5.1.6 Embedded Implementation

It is important to understand that showing the stability of the system ensures convergence of state to the desired state (position in the position stabilization objective) under practical conditions modeled in the system representation. While a lot of analyses is needed for showing the stability, the choice of control law for inputs is to have minimal computations. Now, while implementing the control law, it is worth noting the required inputs and computations. For example, for the implementation of the control law developed in Section 5.1.5 for the inputs u_α and u_β given by (5.71) and (5.60) respectively, the controller block diagram is shown in Figure 5.12.

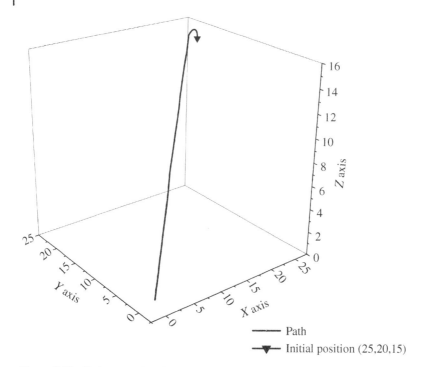

Figure 5.11 Trajectory taken by a unicycle robot in 3D using proposed control law

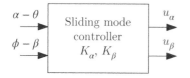

Figure 5.12 Block diagram of embedded sliding-mode controller for 3D position stabilization

Given the computation required for computing the control inputs u_α and u_β as only the signum function, any embedded platform can easily implement the developed controller. This kind of *Top-Down* approach is independent of the selected embedded platform and avoids limitations of embedded implementation.

Section 5.2 further extends the examples presented in this section to switched nonlinear system for a multi-agent application of swarm aggregation (Shah and Vachhani, 2019) as briefly explained in Section 3.5.

5.2 Switched Nonlinear System

A switched nonlinear system comprises multiple subsystems where the system dynamics switched from one sub-system to another. The switching from one subsystem is triggered by a switching signal. It is not sufficient to show the stability of individual subsystem in a switched system, it is also important to check the stability of the system at the time of switching from one subsystem to another. There are many ways by which the stability of switched system can be shown. A popular categorization for showing stability of switched system is based on the way candidate Lyapunov function is selected (Liberzon, 2003). The choice can be based on selecting

1. Same Lyapunov function for all the subsystems.
2. Multiple Lyapunov functions.

Many a time, finding same Lyapunov function for all the subsystem is not feasible. Hence, option of selecting multiple candidate Lyapunov functions can be explored and the interaction of these Lyapunov function during switching can be investigated. These interactions need to satisfy one or the other constraint to show the stability at the time of switching. Let us now describe a switched system representation.

For a switched system with p subsystems $\{s_1, s_2, \ldots, s_p\}$ and initial subsystem s_1, let $\{t_{s_i}(k)\}$ be increasing sequence of switching times. In other words, the state kth time enters in subsystem s_i for any $i = 1, 2, \ldots, p$ at time $t_{s_i}(k)$. For instance, the time instances at which the state enters the subsystem s_1 are $t_{s_1}(1), t_{s_1}(2), t_{s_1}(3), \ldots$ and that of entering the subsystem s_2 are $t_{s_2}(1), t_{s_2}(2), t_{s_2}(3), \ldots$. It is not necessary that a particular relation exists between the times at which state enters from one subsystem to another. But, for the same subsystem $t_{s_i}(k) < t_{s_i}(k+1)$. Further, the time instances when subsystem s_i is active is $\mathcal{I}(s_i) = \bigcup_k \left[t_{s_i}(k), \ t_{s_q}(k+1) \right)$, where $q \neq i$ and the system switched from s_i to s_q after time instance $t_{s_i}(k)$.

Let the dynamics of subsystem s_i be represented as $\dot{x} = f_i(x(t))$, $t \in \mathcal{I}(s_i)$ and V_i be the corresponding candidate function using which it is shown that each subsystem q is Lyapunov stable. It is clear that if the Lyapunov functions V_i for each $i = 1, 2, \ldots, p$ are such that $V_1 = V_2 = \cdots = V_p$, then it is sufficient to claim switched system stability. But, if $V_i \neq V_j$ for any $i \neq j$, then switching from a subsystem to another, if the corresponding Lyapunov function changes needs special attention. It would require a few more conditions to show the switched system stability. One such condition is to check the value of Lyapunov function at the time of entering a subsystem and ensure that it is lesser value in subsequent entry time. In particular, $V_i(t_{s_i}(k+1)) < V_i(t_{s_i}(k))$ for each k as shown in Figure 5.13. The illustration in Figure 5.13 shows a Lyapunov function described for the subsystem s_i where time for entering the subsystem are marked by $t_{s_i}(1), t_{s_i}(2), \ldots$. It shows that the values of Lyapunov function V_{s_i} at the subsequent instances of entry are decreasing. This decreasing trend is applicable to each entry point of a subsystem as well as applicable to each subsystem.

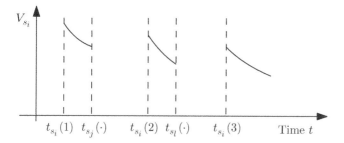

Figure 5.13 Stability condition on multiple Lyapunov functions

An interesting extension is relaxing the condition in the Lyapunov function value in subsequent entry times, as in the case when the value of Lyapunov function at an entry point can be equal as well to that at the previous entry point for the same subsystem, or $V_i(t_{s_i}(k + 1)) \leq V_i(t_{s_i}(k))$. The stability condition can be understood in a simplified manner for a switched system with two subsystems s_1 and s_2. Consider a hypothetical boundary case of having same value of Lyapunov function at each entry point of subsystem s_2 as shown in Figure 5.14.

According to Lu and Brown (2010), the switched system is asymptotically stable if following conditions are satisfied,

1. The system always switches from s_1 to s_2 and remains in s_2 after the last switching.
2. The state in subsystem s_1 is bounded by a value dependent on the immediate precedent entry point in subsystem s_2. In other words the condition is stated as follows: there exists a positive constant m, such that $|V_{s_1}(t)| \leq m|V_{s_2}(t_{s_2}(k))|$, for $t_{s_1}(k + 1) \leq t < t_{s_2}(k + 1)$ and any k.

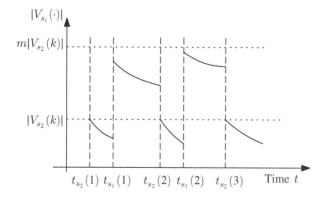

Figure 5.14 Stability condition on multiple Lyapunov functions when the value can be same at entry points

In order to ensure stability of a system for which the control law is derived using algorithmic approach of defining cases and respective control laws, the switched system representation is helpful. Using the concepts of switched system stability and the control law derived in previous section, an interesting embedded controller can be developed for a swarm of unicycle vehicles. The switched system concept for a swarm application is presented next.

5.2.1 Swarm Aggregation as a Switched Nonlinear System

In the swarm aggregation application, a group of multiple agents (unicycle vehicle for the discussion in this book) moves toward a common goal location. When this group moves toward a common goal, the objective is to design a control law that can be implemented on each agent and has a common form. The control law must ensure reaching the goal location avoiding collision with other agents and obstacles in the environment. A path planning approach avoiding the obstacles based on attractive and repulsive field functions can be used for designing control law for each agent. However, when the control system stability is shown, the description using switched system can be explored, particularly when the mobile robots (vehicles) are going to come in contact with each other frequently, as in the case of swarm aggregation. The system will then switch from one control law (attractive) to another (combination of attractive and repulsive when a collision is to be avoided as well).

Now, an approach where cases in algorithmic approach as a switched system are presented. The cases in the swarm aggregation problem now can be described as

1. *Free subsystem*: The control law has only attractive field function and does not require collision avoidance as no obstacle or other agent is nearby.
2. *Engaged subsystem*: The control law has two components attractive as well as repulsive field functions when collision with nearby agents or obstacles is also needed.

In order to apply these cases or switch from one subsystem to another, the triggering signal is to be defined. This triggering signal can be described based on the information about the nearby obstacles/robots obtained from the sensor mounted on the mobile robot/vehicle. For example, the triggering signal switches from free subsystem to engaged subsystem as soon as the sensor detects the presence of an obstacle in its field-of-view. Likewise, it switches back to free subsystem from engaged subsystem as soon as there is the entire field-of-view is clear (no obstacle in the field-of-view). Let us now develop the control laws and respective Lyapunov functions are developed for 2D unicycle vehicle as a mobile agent/robot in swarm aggregation application. Referring to (5.48) for single robot, let the corresponding state vector of ith robot in the swarm be referred by

$X_i = [R_i, \theta_i, \alpha_i]$ and corresponding input be $U_i = [v_i, \omega_i]$. In describing the state vector for ith robot, (R_i, θ_i) describes the position with respect to the goal location and α_i is orientation with respect to inertial reference. The control inputs to ith robot are linear velocity v_i and angular velocity ω_i.

5.2.1.1 Free Subsystem s_1

The control law for ith robot/vehicle implements an attractive field function where the input to the control law is based on aligning itself toward the goal position as given by (5.42). Hence,

$$U_i = \begin{bmatrix} \overline{v}_i \\ -K_i \, \text{sgn}(\alpha_i - \theta_i) \end{bmatrix} \tag{5.76}$$

where $\overline{v}_i$ is a constant positive nominal velocity and K_i is a positive free parameter such that

$$K_i > \frac{\overline{v}_i}{R_c} \tag{5.77}$$

and R_c is the radius of the circle encircled at the goal position where each robot is guaranteed to reach. It is clear that the Lyapunov function to describe the stability of the free subsystem can be borrowed from (5.41). In particular,

$$V_{s_1}^i = \frac{1}{2}(\alpha_i - \theta_i)^2 \tag{5.78}$$

It is worth noting the resemblance of attractive function construct in (3.8) with the Lyapunov function given by (5.78). The objective in case of attractive function given by (3.8) is to reach to the goal which can be considered as a candidate Lyapunov function as well and develop a control law. While the objective in the Lyapunov function given by (5.78) is to align the orientation of the robot toward line-of-sight vector as explained in Section 5.1.4. With the approach suggested in this book, there are variety of ways a control law can be designed by selecting an appropriate candidate Lyapunuv function.

5.2.1.2 Engaged Subsystem s_2

Consider the repulsive function given by (3.9) and observe the motivation behind selecting this function. It collects the distance with respect to the nearest obstacle (or robot in a swarm). The gradient of repulsive function decides the direction in which the robot must move to avoid the collision. Let us consider the motivation behind selecting the repulsive function that moves the robot away from the direction of collision for designing control law in the engaged subsystem. Further the objective for embedded controller design is minimal sensing and processing.

Collating the objectives of minimal sensing and requirements of engaged subsystem, let the sensing requirement be limited to detecting the presence of nearby

Figure 5.15 Engaged subsystem: the jth robot is shown closest to the ith robot

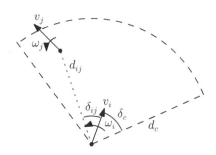

obstacle; once detected then its bearing and distance (inter-agent distance) with respect to the corresponding robot (inter-agent distance). For the ith robot, let the closest detected robot be the jth robot; the inter-agent bearing and distance be δ_{ij} and d_{ij} respectively. Figure 5.15 shows a scenario where jth robot is detected in the field-of-view of the ith robot. The field-of-view of sensor mounted on the ith robot is described by the maximum distance d_c and maximum bearing δ_c that can cover a conical area in front of the robot. It also shows that the ith robot is not in the field-of-view of jth robot. The corresponding velocities are $[v_i, \omega_i]^T$ and $[v_j, \omega_j^T]$ and inter-robot distance and bearing are d_{ij} and δ_{ij}. Notice the construction of measured variable δ_{ij} with respect to the local frame of reference of ith robot.

The control law design is borrowed from Shah and Vachhani (2019). In the engaged subsystem, there is one more measured state which is the inter-agent bearing δ_{ij} and its dynamics is given by

$$\dot{\delta}_{ij} = -\omega_i + \frac{\hat{v}_{ji}}{d_{ij}} \tag{5.79}$$

where $\hat{v}_{ji}$ is the projection of relative velocity of agent j with respect to i ($v_{ji} = v_j - v_i$) perpendicular to the line joining them. The control law for each robot i is given by

$$\omega_i = \kappa_i(\text{sgn}(\dot{\theta}_i) - \text{sgn}(\delta_{ij})) \\ v_i = \max\{v_i(t^i_{s_2}) - (t - t^i_{s_2})\lambda_i, 0\} \tag{5.80}$$

where $t^i_{s_2}$ is the time at which the agent i enters in engaged subsystem s_2, $v_i(t^i_{s_2})$ is the velocity of agent i at $t^i_{s_2}$ and κ_i, λ_i are positive free parameters. Let $\bar{v}$ be $\max_i\{\bar{v}_i\}$ over all agents. This control law ensures collision avoidance and steers the agent out of the engaged subsystem when

$$\kappa_i > \frac{\bar{v}_i}{R_c}s \tag{5.81}$$

$$\lambda_i > \frac{\bar{v}^2}{d_c} \tag{5.82}$$

The derivation of control law is involved and interested readers can refer to Shah and Vachhani (2019). The interesting idea is to capture the formulation of Lyapunov function for the engaged subsystem s_2, which is given by L the sum of attractive and repulsive potentials corresponding to the tasks of aggregation and collision avoidance, respectively. Thus,

$$V_{s_2}^i = V_{att}^i + \gamma_i V_{rep}^i \tag{5.83}$$

with $\gamma_i > 0$ and

$$\begin{aligned} V_{att}^i &= 1 + \cos(\alpha_i - \theta_i) \\ V_{rep}^i &= \cos(\delta_{ij}) - \cos(\delta_c) \end{aligned} \tag{5.84}$$

This result shows that every agent in the swarm is asymptotically stable and falls under one of two categories: (i) robots that reach the target, and (ii) exploratory robots that are on the lookout for a path to the target since robots in category 1 have made the target inaccessible. In this regard, the system observes a state minimizing the average distance to target, with category 2 robots continuously exploring their surroundings. A perturbation to the system may alter the robot category from 1 to 2 or vice versa and allow other exploratory agents to reach the target. In general, such a perturbation can cause the system to settle in a new state that minimizes average distance to target.

5.2.2 Embedded Implementation

The embedded implementation is now straightforward and implements control laws given by (5.76) and (5.80) with the help of a switching signal. The switching signal further is obtained by sensing the presence of obstacle in the field-of-view of sensor mounted on a robot. For the ith robot, the control law implementation can be explained using Figure 5.16.

When a sensor (typically range sensor) gives a reading less than its maximum range reading d_c, the presence of obstacle is detected. Hence, depending on the condition $d_{ij} < d_c$, the presence of obstacle/robot can be detected for each sensor reading. If an obstacle or a robot is detected in the field-of-view of ith robot, then the control law corresponding to the engaged subsystem is applicable otherwise that of free subsystem is applicable. The engaged subsystem control law takes one more input of δ_{ij} as compared to that of free subsystem. It is clear that the control laws in free as well as engaged subsystem do not require computations that are expensive to implement in an embedded platform. Therefore, an example of using switched system for ensuring stability for swarm aggregation using the top-down approach explained in this section can be implemented in any embedded platform.

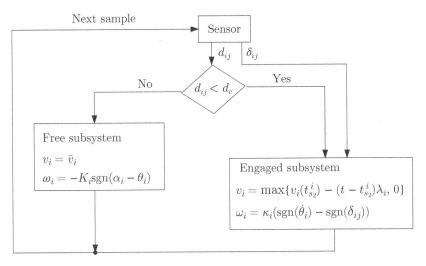

Figure 5.16 Block diagram of embedded controller for swarm aggregation application

5.3 Notes and Further Readings

This chapter introduces robust control strategies for implementation on mobile robots. The introduction and basic definitions give a glimpse of system representation and the stability notions used in control theory. For a detailed treatment of the subject one may refer to specific books in control theory. A comprehensive content on linear systems theory and control design is given in Kailath (1998). For a detailed treatment of nonlinear systems and control, one may refer to Khalil (2002). SMC and their applications are detailed in Edwards and Spurgeon (1998). The position stabilization of vehicle in 2D in Section 5.1.4 is related to the work in Arunkumar et al. (2018). The article details the algorithm in a simpler framework and doesn't resort to sliding-mode theory for the proof. It also includes an application of the algorithm in visual homing of mobile ground robot using panoramic images. Similarly, a detailed analysis of the position stabilization for robots in 3D using SMC can be referred to in Arunkumar and Vachhani (2019).

6

Generic FPGA Architecture Design

The Field Programmable Gate Array (FPGA) is a powerful reconfigurable embedded platform for parallel processing with lots of Input/Output (I/O) pins. The FPGA has gained popularity in robotics, mainly because of the possibilities of implementing parallel independent control loops and parallel processing of multiple sensors. The FPGA technique is a good alternative to Application-Specific Integrated Chip (ASIC). It provides flexibility, shorter design cycle, deterministic time delays, and parallelism that are suitable for a control industry. This chapter is dedicated to understanding the FPGA concepts that are important for efficient implementation.

Another main aspect covered in this chapter is a generic methodology to develop FPGA architecture for a control design. A control design can be implemented in an FPGA by converting a high-level code to FPGA code. However, such an automated methodology adds overheads that are sometimes more than the actual code length occupying extra FPGA area and has the drawback of not exploiting the parallel processing completely. A case study on designing FPGA architecture for low-level controller design for a DC motor and encoder subsystem for a typical robot is also explained as an example of generic control design methodology.

6.1 FPGA Basics and Verilog

An FPGA is a reprogrammable logic device. The building block of an FPGA is a *logic block* that contains *Look-Up Tables* (*LUTs*) and *memory element*. Each LUT is programmed to perform a combinational logic. The memory elements in a logic block give the possibility of implementing sequential digital logic. Since the hardware is not fixed and can be implemented by programming the FPGA, the FPGA is also called as a reconfigurable logic device. An FPGA supports parallelism because hardware can be developed to perform parallel processing and can be configured

Embedded Control for Mobile Robotic Applications, First Edition.
Leena Vachhani, Pranjal Vyas, and Arunkumar G. K.
© 2022 The Institute of Electrical and Electronics Engineers, Inc. Published 2022 by John Wiley & Sons, Inc.
Companion website: www.wiley.com/go/vachhani/embeddedcontrolforroboticapp

on FPGA. The FPGA device also has programmable multiple I/O pins. Each I/O pin can be programmed as input only, output only, or I/O pin. These features of FPGA, viz. reconfiguration capability, possibility of implementing parallel logics and multiple I/Os, are desirable for implementing an embedded controller.

A *Hardware Descriptive Language* (HDL) is used to configure the FPGA. Application softwares are available to plan the placement of logic gates and decide the layout of the architecture. The designer designs the digital architecture and codes it using an HDL. Examples of HDL are Verilog, VHDL, SpectreHDL, and HDL-A. This chapter concentrates on Verilog HDL as its constructs are same as that of "C" language. The main difference between HDL and "C" programming is that the HDL codes are executed in parallel, whereas the "C" codes are executed in sequential manner. The hardware architecture is described using concurrent HDL codes. The functionality of the architecture can be described using combination of following common modeling techniques:

- *Behavioral modeling*: describes the behavior of the architecture.
- *Structural modeling*: describes the interconnection between different sub-modules.

Basic constructs in Verilog are register (defined by *reg*) and wire (defined by *wire*). The register is a storage element and wire carries the signal without storing it. The wires are used for interconnecting the sub-modules. The inputs are already latched, whereas the outputs in behavioral modeling need to be stored in a register. The register and wire have similarity with "C" language variables. If the variable is being assigned (it appears on the Left-Hand Side (LHS) of the assignment statement), it needs to be declared as register in HDL. If a particular variable is always used for reading purpose (it appears on the Right-Hand Side [RHS] of the assignment statement), then it needs to be declared as "wire" for that particular module. An example of a Verilog coding for full adder is illustrated next.

```
Verilog code for full adder:
  reg sum,carry;
    wire a,b;
always @(a,b)
  begin
  sum <= a xor b;
  carry <= a and b;
end
```

The "always" construct used in this example triggers the execution of a block (bounded by "begin" and "end") whenever there is an event on the signals "a" and "b." The list of events is called sensitivity list. The sensitivity list for a combinational circuit contains all the signals that appear on the RHS of the assignment statement and the ones that are used in conditional statements.

Common constructs in a digital architecture are classified into two categories, viz. combinational and sequential circuits. A combination circuit implements a boolean logic which depends only on the present input, whereas a sequential circuit implements a boolean logic that depends not only on the present input, but also the past input and/or output. The sequential circuit has memory and operates at clock events. Common combinational circuits are *multiplexer, demultiplexer, adder/subtractor, comparator, accumulator, decoder, encoder, arithmetic and logical unit*, etc. and common sequential circuits are *counter, timer, flip-flop, register, memory, shifter, state machine*, etc. As illustrative examples of Verilog sample codes, let us review 8×1 multiplexer, 4-bit counter with synchronous and asynchronous reset shown in Figure 6.1.

The Verilog sample code for 8×1 multiplexer which has C language-like constructs, is as follows:

```
Verilog code for 8 × 1 multiplexer:
module mux(in,sel,out);
    input [7:0] in;
    input [2:0] sel;
    output reg out;
    always @(in, sel)
    begin
      case(sel)
          3'd0: out <= in[0];
          3'd1: out <= in[1];
          3'd2: out <= in[2];
          3'd3: out <= in[3];
          3'd4: out <= in[4];
```

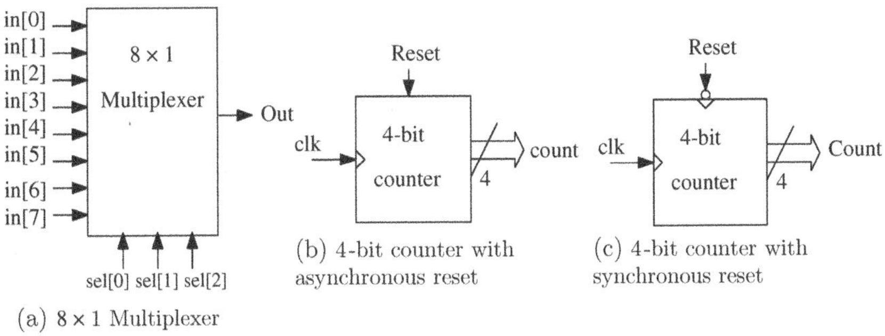

(a) 8×1 Multiplexer

(b) 4-bit counter with asynchronous reset

(c) 4-bit counter with synchronous reset

Figure 6.1 Illustration of combinational and sequential circuits

```
            3'd5: out <= in[5];
            3'd6: out <= in[6];
            3'd7: out <= in[7];
        endcase
    end
endmodule
```

It is worth noting the difference between C language code and Verilog code constructs. The difference as stated earlier is the sequential and concurrent executions of statements. Each statement inside the "always" block is executed in parallel when the event is triggered in its sensitivity list. In particular, whenever there is a change in signals in the sensitivity list, the event is triggered and statements inside the "always" block are executed in parallel. In the example code of 8 × 1 multiplexer, the sensitivity list has two input signals "in" and "sel." There is only one statement which is the "case" statement. There are multiple choices based on the input signal value "sel." Thereby, the multiplexer output value in register (single bit) is changed either when the input signal "in" changes or the selection choice via the input signal "sel" is changed.

The next example demonstrates the bus assignment as an array using an example Verilog code for a 4-bit counter.

```
Verilog code for 4-bit counter with synchronous reset:
module counter(clk, reset, count);
    input clk;
    input reset;
    output [3:0] count;
    reg [3:0] count;
    always @(posedge clk)
    begin
      if (reset==1'b0)
        count <= 4'd0;
      else
        count <=count+1;
    end
  endmodule
```

Note the use of edge-triggering events in the sensitivity list. This usage allows the implementation of sequential circuits. Sequential circuit with an asynchronous reset (negative edge triggered) sequential circuit is coded as follows:

```
Verilog code for 4-bit counter with asynchronous reset:
module  counter(clk, reset, count);
```

```
      input clk;
      input reset;
      output [3:0] count;
      reg [3:0] count;
      always @(posedge clk, negedge reset)
      begin
        if (reset==1'b0)
          count <= 4'd0;
        else
          count <= count+1;
      end
  endmodule
```

Note that the sensitivity list now contains the negative edge-triggering of reset signal as well. A module can have multiple submodules and the approach is called structural modeling. The submodules are interconnected by instantiating a submodule and maintaining the order of I/O list or specifically describing the I/O list. An example of 16-bit counter using four 4-bit counters as shown in Figure 6.2 is coded as follows:

```
Verilog code for 16-bit counter:
module  sixteen_counter(clock, rst, count);
   input clk;
     input rst;
     output [15:0] count;
counter cn1(clock, rst, count[3:0]);
counter cn2(count[3], rst, count[7:4]);
counter cn3(.clk(count[7]),.reset(rst),.count(count[11:8]));
counter cn4(.clk(count[11]),.reset(rst),.count(count[15:12]));
endmodule
```

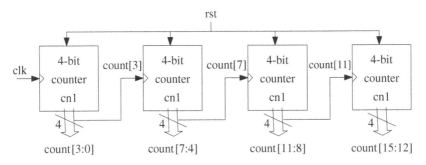

Figure 6.2 Concept of using submodules

The code shows two ways of giving assignments. In the submodules cn1 and cn2, the assignments are based on the sequence of signals defined in the counter module. In particular, the sequence of signals in counter module is clock, reset, and 4-bit count. Hence, the assignments in cn1 and cn2 follow this sequence. In the submodules cn3 and cn4, the assignments are explicit using the dot followed by signal name in the module. For example, the .clk(count[7]) assigns signal count[7] to the clk of module cn3. Further, note the assignment of signal "count[3]" to "clk" in counter "cn2" and similar assignments of Most Significant Bit (MSB) of lower bit counters to the next higher-up to clock. An HDL code can have combinations of modules defined in structural and behavioral modeling.

The hardware architecture for controller implementation would need combinational and sequential circuit designs. One of the approaches is to consider the design of *Finite State Machine (FSM)*. An FSM captures all the states (finite number of states) of a machine/automata. Each state typically describes the values of output variable or any internal variable. The state is triggered on the action of a control input. Consider an example of designing circuit for two 1-bit output variables y_1 and y_2. Since there exist four combinations of values for the two variables, there are four states in the design as shown in Figure 6.3. In the illustration of this 4-state FSM, the FSM remains in the same state if the input variable is 0 and triggers to the next state if the input variable is 1.

The FSM design is obtained in the context of controller implementation in this chapter. In other words, the FSM is designed for internal operations happening within a sample time T_s of control loop as shown in Figure 6.4.

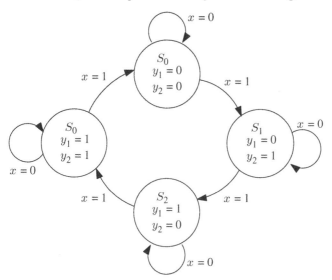

Figure 6.3 An illustration of 4-state FSM

Figure 6.4 Finite state machine (FSM) as a black box

Section 6.2 presents an approach for designing the architecture using basic combinational and sequential blocks.

6.2 Systematic Approach for Designing Architecture Using FSM

The architecture consists of a *control logic* and *datapath*. The *control logic* is responsible for generating control signal at the correct time instant, whereas the *datapath* is responsible for data transfer and storage. The first step of the architecture design is representing the functionality of the controller as *FSM* (Givargis and Vahid, 2012). The FSM is a state machine with finite number of states. The FSM considers that the controller can be in only one state at a given time and all the states are represented in FSM. The controller may change its state with the triggering of an event. The procedure for designing architecture using FSM[1] is explained using following steps:

- *Step 1*: Create separate registers for all the declared variables. The load input of each register is the control output of the control logic. Different inputs to a register are identified by finding the assignment statements for which the register variable is on the LHS of the assignment. Multiple inputs to a register are connected through a multiplexer. The select input of the multiplexer is also the control output.
- *Step 2*: Create a functional unit for each arithmetic/logical operation in state machine.
- *Step 3*: Connect registers and functional unit. The RHS of assignment and condition statements provide the information on input to the functional unit.
- *Step 4*: Create a unique identifier for each control input and output of the datapath.
- *Step 5*: Design control logic with these unique identifiers. Input to the datapath is the output of the control logic and vice versa.

The architecture design is further explained using an example of designing PID controller in Section 6.2.1.

---

1 [1]/with permission of Elsevier.

6.2.1 PID Controller Architecture

The difference equation for conventional Proportional Integrator Differentiator (PID) controller given by (2.7) is rearranged as follows:

$$u[n] = u[n-1] + \left[T_1 e[n] + T_2 e[n-1] + T_3 e[n-2] \right] \tag{6.1}$$

where $T_1 = K_p + K_i \frac{T_s}{2} + \frac{K_d}{T_s}$, $T_2 = -K_p + K_i \frac{T}{2} - 2\frac{K_d}{T_s}$, $T_3 = \frac{K_d}{T_s}$, T_s is the sampling time, K_p, K_i, and K_d are proportional, integral, and differentiation gains, respectively. Let there be an architecture constraint of availability of a single adder and multiplier.

The flow chart for the execution of (6.1) is shown in Figure 6.5. The functions in each process box of the flow chart are executed in parallel. The execution of the additions is shown in separate process box to satisfy the constraint of single adder and multiplier. The *step 1* of the systematic procedure identifies the declared

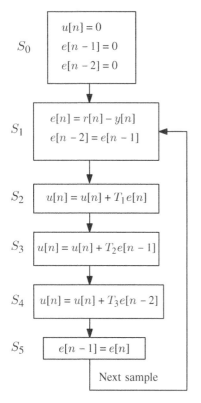

Figure 6.5 Flow chart for PID controller

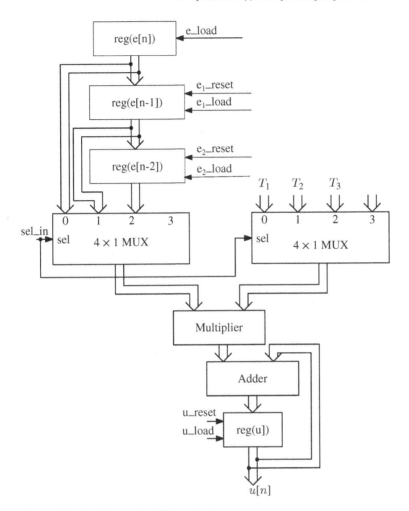

Figure 6.6 Architecture for PID controller

variables from the flow chart. These are the variables on the LHS of the assignment statements. For the flow chart shown in Figure 6.5, the declared variables are $u[n]$, $e[n]$, $e[n-1]$, and $e[n-2]$. The architecture for PID controller will have registers associated with these declared variables as shown in Figure 6.6.

In the *step 2*, the functional units are worked out. The functional units identified for PID controller are adder and multiplier. The RHS of assignment statement and condition statements provide the requirement for the types of functional unit.

The *step 3* connects the functional units and registers. The required multiplications are $T_1 * e[n]$, $T_2 * e[n-1]$, and $T_3 * e[n-2]$. Therefore, one input to the multiplier is either $e[n]$, $e[n-1]$, or $e[n-2]$ and the other input to the multiplier is either T_1, T_2, or T_3. A 4×1 MUX is used at each input of the multiplier. Both the MUXs have corresponding inputs. For example, the input lines, that are selected when sel_in is "0," for each MUX are $e[n]$ and T_1, respectively. The output of the multiplier is always one input to the adder. The other input to the adder is always $u[n]$. The LHS of the assignment performing addition is always $u[n]$; therefore, output of the adder is stored in register $u[n]$. The registers $e[n]$, $e[n-1]$, and $e[n-2]$ are receiving input from a single source; therefore, their inputs are directly connected with their respective source.

The new value in the respective register is stored at a particular time instant. Unique identifiers for control signals are decided. The control signals in the PID architecture are e_load, e_1_load, e_1_reset, e_2_load, e_2_reset, sel_in, u_reset, and u_load.

In the next step, the FSM for the design is worked out and is shown in Figure 6.7. Each state of the FSM corresponds to a process box in the flowchart. The control signals that are modified in a state are indicated inside the circle. The condition that triggers the state change is indicated over the directed edge between two states. The state transits to a new state if the condition is met at the clock event. The design of control logic takes help of this FSM. The inputs to the control logic are enable and next sample event. Outputs of the control logic are all the control signals as shown in Figure 6.8. The control logic uses a register for storing the present state of the controller. Since the FSM has five states, a 3-bit register is used with its inputs and outputs termed as I_i and O_i, $\forall i = 0, 1, 2$, respectively. This register loads input at every clock instant.

The truth table for the control logic is shown in Table 6.1. Each row in the truth table represents an edge in the FSM. For example, first row represents state S_0 with the directed edge with enable="0." The column entries for this row have input enable as "0" and sample as "X" (don't care). The output remains at state S_0 with u_reset, e_1_reset, and e_2_reset as logic "1" and rest output as don't care. Similarly, second row represents state S_0 with enable="1" that triggers to state S_1 where outputs (u_reset, e_1_reset, and e_2_reset) that were activated in previous state (S_0) are deactivated (logic "0") and outputs e_load and e_2_load are activated (logic "1").

Control logic is a combinational logic based on the truth table. Next, the design of a sliding-mode controller is explained as an illustration.

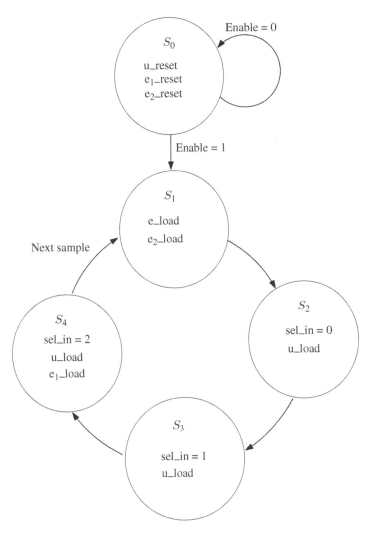

Figure 6.7 State machine for PID controller

6.2.2 Sliding-Mode Controller Architecture

Let a nonlinear dynamic system be described by

$$\dot{x}(t) = f(x, t) + B(x, t)u(t) \tag{6.2}$$

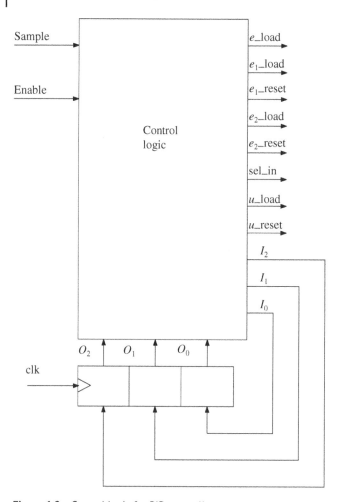

Figure 6.8 Control logic for PID controller

Let the switching function described in Sabanovic et al. (2004) for the controller be selected as the linear combination $\sigma(x) \triangleq s_1 x_1 + s_2 x_2 + \cdots + s_n x_n$, where weights, $s_i \geq 0, \forall i = 1, 2, \ldots, n$. The trajectories are forced to slide along the surface $\dot{\sigma}(x) = 0$. The sliding-mode control always switches from one state to another based on the sign of switching function. Let the system described by (6.2) have single input. A typical sliding-mode control-law for single input has the form

$$u(x(t)) = \begin{cases} u^+(x) & \text{if } \sigma(x) > 0 \\ u^-(x) & \text{if } \sigma(x) < 0 \end{cases} \tag{6.3}$$

Table 6.1 Truth table for PID control logic

Inputs					Outputs										
Enable	Sample	O_2	O_1	O_0	e_load	e_1_load	e_1_reset	e_2_load	e_2_reset	sel_in	u_load	u_reset	I_2	I_1	I_0
0	X	0	0	0	X	X	1	X	1	X	X	1	0	0	0
1	X	0	0	0	1	0	0	1	0	X	0	0	0	0	1
X	X	0	0	1	0	0	0	0	0	0	1	0	0	1	0
X	X	0	1	0	0	0	0	0	0	1	1	0	0	1	1
X	X	0	1	1	0	0	0	0	0	2	1	0	1	0	0
X	0	1	0	0	0	0	0	0	0	X	0	0	1	0	0
X	1	1	0	0	1	0	0	1	0	X	0	0	0	0	1

Source: Vachhani (2013).

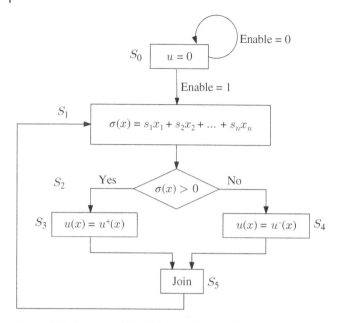

Figure 6.9 Flow chart for sliding-mode controller

The design steps for developing architecture to generate sliding-mode control law $u(x(t))$ are as follows: The flow chart for the functionality of the control law is shown in Figure 6.9. Assume that the a multiplier and adder with multiple inputs is available (The architecture for multiple inputs with one multiplier and adder block has been designed in Section 6.2.1). This block represents functional block for calculation of $\sigma(x)$. The *step 1* of the design procedure finds declared variables from the flow chart. In the flow chart shown in Figure 6.9, there are two declared variables viz. $\sigma(x)$ and $u(x)$. Therefore, two registers for each variable are considered in the architecture as shown in Figure 6.10. The *step 2* identifies the functional units. The functional units for sliding-mode controller are required for the calculations of $\sigma(x)$, $u^+(x)$, and $u^-(x)$ and comparison of $\sigma(x)$ with 0. The logical unit for

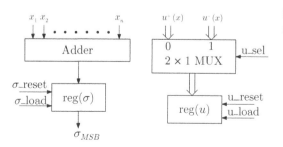

Figure 6.10 Architecture for sliding-mode controller

Figure 6.11 State Machine for
sliding-mode controller

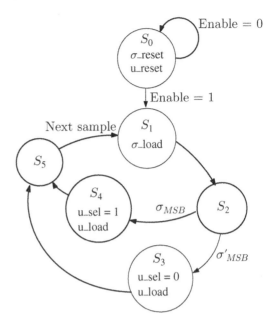

Figure 6.11 State Machine for sliding-mode controller

comparison can be avoided because the comparison is with respect to 0 and can be obtained by the MSB of register σ (the MSB signifies sign).

For connecting functional units and the registers in *step 3*, the inputs to the registers are identified. The input to the register σ is always the function calculated for $\sigma(x)$. Therefore, the input connection to register σ is the output of functional unit that calculates $\sigma(x)$. The input to the register u is either $u^+(x)$ or $u^-(x)$. Therefore, a 2×1 MUX is used at the input of the register u.

In the *step 4*, control variables are termed. The control variables for architecture shown in Figure 6.10 are termed as σ_load, σ_reset, u_reset, u_load, and u_reset. The last step of the design procedure is to generate these control signals at the correct time instant. The FSM for sliding-mode control is designed and shown in Figure 6.11. The FSM is implemented as control logic as shown in Figure 6.12. The input–output relation for the control logic design is given by the truth table in Table 6.2. Each entry in the truth table reflects an edge in the FSM.

6.3 FPGA Implementation

Let us understand the benefits that an HDL like Verilog coding provides while implementing the designed architecture for FPGA. The state machine in control unit implements the sequential logic while other combinational logic is implemented as architectural components. A sample Verilog code implementing the

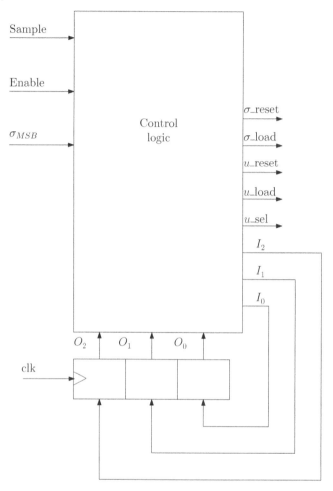

Figure 6.12 Control logic design for sliding-mode controller. Source: Vachhani (2013).

control unit shown in Figure 6.8 (combined with the registers for e, e_1, and e_2 of architectural components from Figure 6.6) is as follows:

```
module  control_unit(
    input clk,
    input enable,
    input sample,
    input [31:0] e,
    output reg sel_in,
```

Table 6.2 Truth table for sliding-mode control logic

Inputs						Outputs							
Enable	Sample	σ_{MSB}	O_2	O_1	O_0	σ_reset	σ_load	u_reset	u_load	u_sel	I_2	I_1	I_0
0	X	X	0	0	0	1	X	1	X	X	0	0	0
1	X	X	0	0	0	0	1	0	0	X	0	0	1
X	X	X	0	0	1	0	0	0	0	X	0	1	0
X	X	1	0	1	0	0	0	0	1	1	1	0	0
X	X	0	0	1	0	0	0	0	1	0	0	1	1
X	X	X	0	1	1	0	0	0	0	X	1	0	1
X	X	X	1	0	0	0	0	0	0	X	1	0	1
X	1	X	1	0	1	0	1	0	0	X	0	0	1
X	0	X	1	0	1	0	0	0	0	X	1	0	1

```
output reg u_reset,
output reg u_load,
output reg [31:0] e1,
output reg [31:0] e2);

reg [2:0] state = 3'd0;
always @ (posedge clk)
begin
  case (state)
  3'd0:  begin
      e1=32'd0;
      e2=32'd0;
      u_reset = 1'b1;
      if (enable == 1'b1)
          state = 3'd1;
      end
  3'd1: begin
      e2=e1;
      u_reset = 1'b0;
      state = 3'd2;
      end
  3'd2: begin
      sel_in = 2'd0;
      u_load = 1'b1;
```

```
                    state = 3'd3;
                end
          3'd3: begin
                sel_in = 2'd1;
                u_load = 1'b1;
                state = 3'd4;
                end
          default: begin
                sel_in = 2'd2;
                u_load = 1'b1;
                e1=e;
                if (sample == 1'b1)
                   state = 3'd1;
              end
          endcase
    end
endmodule
```

As explained earlier, module description covers input and output signals. The list of wires are internal signals to the module as we designed the control logic given by Table 6.1. In Figure 6.8, the registers O_2, O_1, and O_0 (or 3'bit variable O) is the "state" register in the Verilog code. Depending on the current value of register "state" the actions are listed using the "case" construct which resembles for a multiplexer architecture.

The list of internal signals are as taken from Table 6.1 are e_load, e_1_load, e_1_reset, e_2_load, e_2_reset, sel_in, u_load, u_reset, I_2, I_1, and I_0. The signals e_load, e_1_load, e_1_reset, e_2_load, and e_2_reset corresponding to registers e, e_1, and e_2 are used for reset and loading the corresponding registers e, e_1, and e_2. These signals are not explicitly generated in the sample Verilog code as the functionality of loading and resetting is incorporated in the corresponding state. The signals I_2, I_1, and I_0 (or 3'bit variable I) represent next state and its implemented as the statement "state = 3'd1," "state = 3'd2," etc. in the corresponding present state. Now, the topmost sample Verilog module implementing the architecture shown in Figure 6.6 is shown using a block diagram in Figure 6.13 (the corresponding Verilog code is available in top_PID.v).

The input–output list of the top module has the input signal e corresponding to error signal of controller and output signal u corresponding to the control signal (input to the system). The sample code considers the PID time constants T_1, T_2, and T_3 tuned to 435, 1023, and 2345 unit values, respectively. It is worth noting the connections of signals from one module to another. The sequence in which they appear describes the connections with the respective input–output list of the

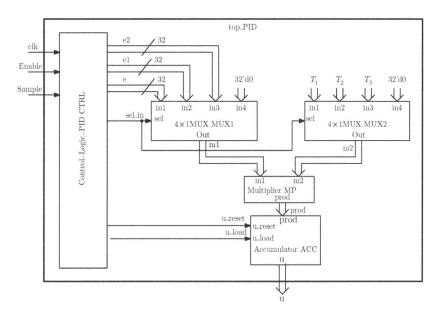

Figure 6.13 Block diagram illustrating topmost Verilog module

corresponding module. For example 32-bit 4×1 multiplexer has four 32-bit inputs, one 2-bit select line, and a 32-bit output. The module name MUX1 has 32-bit inputs e, e_1, e_2, and 0 and 32-bit output denoted by signal m1. This signal m1 then is an input to the multiplier.

The synthesis result on Xilinx ISE tool for example architecture with normal optimization efforts is tabulated in Table 6.3. The FPGA area requirements are then noted. The number of slices, number of slice registers, number of 4-input LUTs, and number of 18×18 multipliers required for the implementation is presented as a row element in Table 6.3. The area consumption is noted for 32-bit input that requires 32×32 multiplier and gives a 64-bit output. The area utilization is 13% slices, 5% slice registers, and 12% 4-input LUTs on XC2S100E FPGA device. Similar results on the area consumption for sliding-mode controller implementation can be obtained that depends on the chosen functions $u^+(x)$ and $u^-(x)$.

Table 6.3 FPGA area consumption for PID controller with 32-bit input

Slices	Slice flip-flops	4-input LUT	18×18 Multiplier
128	103	239	4

It is worth noting the maximum clock frequency for the implementation is 63.105 MHz. In particular, the minimum delay introduced by the controller is 15.84 ns, if the clock frequency of 63.105 MHz is connected to the FPGA-based controller. This delay calculation can further be used for selecting sampling frequency for the control loop. However, typically in the FPGA implementation, computation delay is negligible as compared to the actuator response time. The Section 6.4 further provides another advantage of using FPGA in controller designs by utilizing its parallel processing capabilities.

6.4 Parallel Implementation of Multiple Controllers

We now know that FPGA can be used for real-time computing requirements. There is one more powerful way where use of FPGA as an embedded platform will benefit. The FPGAs facilitate parallel computing and it is not necessary that each module in the parallel architecture must have the same structure. Moreover, a typical FPGA has hundreds of IO pins. Therefore, applications where many control loops are executed in parallel. For example, a vehicle has multiple propellers and multiple sensors. Each propeller will have a local loop to control its commanded speed. Each sensor will have its own processing requirements. Each of the control loop and sensor processing is independent of each other at the lower-level design and can be implemented in parallel (without interacting with each other) without adding any extra processing delays in issuing commands.

In Figure 6.14, a block diagram shows parallel control loops implemented on FPGA interfaced to corresponding sensor and actuator sets. Since FPGA has

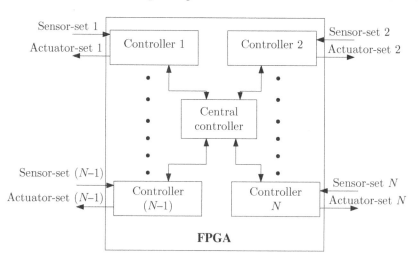

Figure 6.14 Parallel control loops using FPGA

multiple Input–Output (IO) ports or IO Blocks (IOBs) as explained in Section 1.4 using Figure 1.17 it is possible through FPGA architecture development to have parallel controllers with their reference inputs received from a central controller. With the parallel configuration of controller architecture, the inputs through the central controller can be issued in parallel and each controller works in parallel, independent of each other. Note that each controller can implement its own control-law as in PID or sliding-mode or any other kind. No sequential delays in executing controller commands are the main benefit of parallel architecture. It is clear that using the inherent capabilities of embedded platforms, their computation limitations can be easily avoided.

6.5 Notes and Further Readings

This chapter gives a method for controller design in FPGA. There exist other direct methods of implementing architectures in FPGA. Many cross-compilers (such as MATLAB© HDL coder) exist that converts high-level language codes or models in corresponding HDL codes and facilitate direct generic method for synthesizing architectures in FPGA. However, there is additional wrapper HDL codes generated using this generic method which does not provide optimal solution for FPGA area consumption and maximum clock frequency requirements for the hardware architecture designs. If full benefits of FPGA on achieving real-time performance and parallel processing are to be explored, then implementing controller using the design methodology presented in this chapter is advantageous.

For exploring more on pipelining and parallel processing a good read is a book by Parhi Parhi (2007). Also, interested readers may refer to Palnitkar (1996) for learning HDL using Verilog and Givargis and Vahid (2012), Vahid and Lysecky (2007) for learning HDL using VHDL with many worked out examples.

Various controllers on FPGA such as PID controllers along with modern controllers like Model Predictive Controller (MPC), and Fuzzy logic controller have been reported (Kozak, 2012). Further, interesting works exist on FPGA for implementing adaptive backstepping sliding-mode control (Lin et al., 2007), fuzzy logic controller (Calvillo et al., 2011), and model-based predictive control (Vyncke et al., 2013). The interested readers can also refer to work reported in Vyncke et al. (2013) where FPGA capability is explored in effective implementation of the controller. The speed benefit is obtained by parallel computation of prediction and computation steps on different inputs and pipelining the calculation of cost function.

7

Summary

This book covers concepts in three main areas: embedded systems, control theory, and mobile robotic applications. The topics which connect these areas are studied and systematically arranged to establish connection between them. There are many emerging robotic applications that are to be solved using embedded control methods. This book brings out two ways to address embedded control design: (i) for scenarios where the option for selecting embedded platform is available, and (ii) for scenarios where embedded platform is already selected to take care of other interfaces on the mobile robot. In both the scenarios, the limitations of available computing resources are suggested to be considered at the design stage itself. Following the methods studied in this book, embedded controller designs can be developed for emerging technologies on multi-agent applications. Possibility of defining and satisfying new control objectives for emerging technologies built on customized embedded platforms can be explored based on the foundations placed in this book.

Embedded Control for Mobile Robotic Applications, First Edition.
Leena Vachhani, Pranjal Vyas, and Arunkumar G. K.
© 2022 The Institute of Electrical and Electronics Engineers, Inc. Published 2022 by John Wiley & Sons, Inc.
Companion website: www.wiley.com/go/vachhani/embeddedcontrolforroboticapp

Bibliography

S. Aggarwal, P. K. Meher, and K. Khare. Concept, design, and implementation of reconfigurable CORDIC. *IEEE Transactions on Very Large Scale Integration (VLSI) Systems*, 24(4):1588–1592, 2016.

G. K. Arunkumar and L. Vachhani. Autonomous homing in 3D using partial state feedback on a novel sliding surface. In *2019 18th European Control Conference (ECC)*, pages 1–7, 2019.

G. K. Arunkumar, A. Sabnis, and L. Vachhani. Robust steering control for autonomous homing and its application in visual homing under practical conditions. *Journal of Intelligent and Robotic Systems*, 89(3):403–419, 2018. ISSN 1573-0409.

T. Balch and M. Hybinette. Social potentials for scalable multi-robot formations. In *Proceedings 2000 ICRA. Millennium Conference. IEEE International Conference on Robotics and Automation. Symposia Proceedings (Cat. No.00CH37065)*, volume 1, pages 73–80, 2000.

M. G. Bell. Hyperstar: A multi-path Astar algorithm for risk averse vehicle navigation. *Transportation Research Part B: Methodological*, 43(1):97–107, 2009.

D. Biswas and K. Maharatna. A CORDIC-based low-power statistical feature computation engine for WSN applications. *Circuits, Systems, and Signal Processing*, 34(12):4011–4028, 2015. ISSN 1531-5878.

M. Brambilla, E. Ferrante, M. Birattari, and M. Dorigo. Swarm robotics: a review from the swarm engineering perspective. *Swarm Intelligence*, 7(1):1–41, 2013. ISSN 1935-3820.

R. W. Brockett. Asymptotic stability and feedback stabilization. In (eds. R. W. Brockett, R. S. Millman and H. J. Sussmann) *Differential Geometric Control Theory*, volume 27, pages 181–191. Birkhauser: Boston, 1983.

C. F. Calvillo, F. Martell, J. L. Elizondo, A. Ávila, M. E. Macías, M. Rivera, and J. Rodriguez. Rotor current fuzzy control of a DFIG with an indirect matrix converter. In *IECON 2011 - 37th Annual Conference of the IEEE Industrial Electronics Society*, pages 4296–4301, 2011.

Embedded Control for Mobile Robotic Applications, First Edition.
Leena Vachhani, Pranjal Vyas, and Arunkumar G. K.
© 2022 The Institute of Electrical and Electronics Engineers, Inc. Published 2022 by John Wiley & Sons, Inc.
Companion website: www.wiley.com/go/vachhani/embeddedcontrolforroboticapp

S. Camazine, N. R. Franks, J. Sneyd, E. Bonabeau, J.-L. Deneubourg, and G. Theraula. *Self-Organization in Biological Systems*. Princeton University Press, 2001. ISBN 9780691116242.

A. Candra, M. A. Budiman, and K. Hartanto. Dijkstra's and A-Star in finding the shortest path: a tutorial. In *2020 International Conference on Data Science, Artificial Intelligence, and Business Analytics (DATABIA)*, pages 28–32. IEEE, 2020.

G. Caprari, T. Estier, and R. Siegwart. Fascination of down scaling-Alice the sugar cube robot. *Journal of Micro-Mechatronics*, 1(3):177–189, 2001.

A. Chauhan, P. Vyas, L. Vachhani, and A. Maity. Optimal path planning for a non-holonomic robot using interval analysis. In *2018 Indian Control Conference (ICC)*, pages 184–189. IEEE, 2018.

S. Y. Chen. Kalman filter for robot vision: a survey. *IEEE Transactions on Industrial Electronics*, 59(11):4409–4420, 2011.

S. H. Chiew, W. Zhao, and T. H. Go. Swarming coordination with robust control Lyapunov function approach. *Journal of Intelligent and Robotic Systems*, 78(3):499–515, 2015. ISSN 1573-0409.

A. De Luca, G. Oriolo, and M. Vendittelli. *Control of Wheeled Mobile Robots: An Experimental Overview*, pages 181–226. Springer, Berlin, Heidelberg, 2001. ISBN 978-3-540-45000-9.

F. Dellaert, D. Fox, W. Burgard, and S. Thrun. Monte Carlo localization for mobile robots. In *Proceedings 1999 IEEE International Conference on Robotics and Automation*, volume 2, pages 1322–1328. IEEE, 1999.

H. Dong, W. Li, J. Zhu, and S. Duan. The path planning for mobile robot based on Voronoi diagram. In *2010 3rd International Conference on Intelligent Networks and Intelligent Systems*, pages 446–449. IEEE, 2010.

J. W. Durham and F. Bullo. Smooth nearness-diagram navigation. In *2008 IEEE/RSJ International Conference on Intelligent Robots and Systems*, pages 690–695. IEEE, 2008.

C. Edwards and S. K. Spurgeon. *Sliding Mode Control: Theory And Applications, Series in Systems and Control*. CRC Press, 1998. ISBN 9780748406012.

A. Farinelli and L. Iocchi. Planning trajectories in dynamic environments using a gradient method. In *Proc. of the International RoboCup Symposium 2003*, pages 320–331. Springer, 2003.

T. I. Fossen. *Guidance and Control of Ocean Vehicles*. Wiley, 1994. ISBN 9780471941132.

D. Fox, W. Burgard, and S. Thrun. The dynamic window approach to collision avoidance. *IEEE Robotics and Automation Magazine*, 4(1):23–33, 1997.

D. Fox, W. Burgard, and S. Thrun. Active Markov localization for mobile robots. *Robotics and Autonomous Systems*, 25(3–4):195–207, 1999.

D. Fox, W. Burgard, F. Dellaert, and S. Thrun. Monte Carlo localization: efficient position estimation for mobile robots. *AAAI/IAAI*, 1999(343–349):2, 1999a.

D. Fox, W. Burgard, and S. Thrun. Markov localization for mobile robots in dynamic environments. *Journal of Artificial Intelligence Research*, 11:391–427, 1999b.

J. Fuentes-Pacheco, J. Ruiz-Ascencio, and J. M. Rendón-Mancha. Visual simultaneous localization and mapping: a survey. *Artificial Intelligence Review*, 43(1):55–81, 2015.

G. J. García, C. A. Jara, J. Pomares, A. Alabdo, L. M. Poggi, and F. Torres. A survey on FPGA-based sensor systems: towards intelligent and reconfigurable low-power sensors for computer vision, control and signal processing. *Sensors*, 14(4):6247–6278, 2014.

M. Garrido, P. Källström, M. Kumm, and O. Gustafsson. CORDIC II: A new improved CORDIC algorithm. *IEEE Transactions on Circuits and Systems II: Express Briefs*, 63(2):186–190, 2016.

A. Gasparetto, P. Boscariol, A. Lanzutti, and R. Vidoni. Trajectory planning in robotics. *Mathematics in Computer Science*, 6(3):269–279, 2012.

T. Givargis and F. Vahid. *Embedded System Design: A Unified Hardware / Software Introduction*. John Wiley & Sons, 2012.

P. Glotfelter and M. Egerstedt. A parametric MPC approach to balancing the cost of abstraction for differential-drive mobile robots. In *2018 IEEE International Conference on Robotics and Automation (ICRA)*, pages 732–737, 2018.

Y. K. Hwang, N. Ahuja, et al. A potential field approach to path planning. *IEEE Transactions on Robotics and Automation*, 8(1):23–32, 1992.

F. Islam, Ja. Nasir, U. Malik, Y. Ayaz, and O. Hasan. RRT*-Smart: Rapid convergence implementation of RRT* towards optimal solution. In *2012 IEEE International Conference on Mechatronics and Automation*, pages 1651–1656. IEEE, 2012.

F. J. Jaime, M. A. Sanchez, J. Hormigo, J. Villalba, and E. L. Zapata. Enhanced scaling-free CORDIC. *IEEE Transactions on Circuits and Systems I: Regular Papers*, 57(7):1654–1662, 2010.

L. Jaulin. Robust set-membership state estimation; application to underwater robotics. *Automatica*, 45(1):202–206, 2009.

L. Jaulin and F. Le Bars. An interval approach for stability analysis: application to sailboat robotics. *IEEE Transactions on Robotics*, 29(1):282–287, 2012.

L. Jaulin, M. Kieffer, O. Didrit, and E. Walter. *Applied Interval Analysis*. Springer-Verlag London Ltd., London, 2001.

L. Jaulin, A. Caiti, M. Carreras, V. Creuze, F. Plumet, B. Zerr, and A. Billon-Coat. *Marine Robotics and Applications*, volume 10. Springer, 2018.

L. Jetto, S. Longhi, and G. Venturini. Development and experimental validation of an adaptive extended Kalman filter for the localization of mobile robots. *IEEE Transactions on Robotics and Automation*, 15(2):219–229, 1999.

B. Johann and Y. Koren. The vector field histogram-fast obstacle avoidance for mobile robots. *IEEE Transactions on Robotics and Automation*, 7(3):278–288, 1991.

T. Kailath. *Linear Systems, Prentice-Hall Information and System Sciences Series.* Prentice Hall International, 1998. ISBN 9789814024785.

S. Karaman, M. R. Walter, A. Perez, E. Frazzoli, and S. Teller. Anytime motion planning using the RRT. In *2011 IEEE International Conference on Robotics and Automation*, pages 1478–1483. IEEE, 2011.

B. I. Kazem, H. H. Ali, and M. M. Mustafa. Modified vector field histogram with a neural network learning model for mobile robot path planning and obstacle avoidance. *International Journal of Advanced Computer Technology*, 2(5):166–173, 2010.

H. K. Khalil. *Nonlinear Systems.* Pearson Education, Prentice Hall, 2002. ISBN 9780130673893.

K. Konolige. A gradient method for realtime robot control. In *Proceedings. 2000 IEEE/RSJ International Conference on Intelligent Robots and Systems (IROS 2000)(Cat. No. 00CH37113)*, volume 1, pages 639–646. IEEE, 2000.

S. Kozak. Advanced control engineering methods in modern technological applications. In *Proceedings of the 13th International Carpathian Control Conference (ICCC)*, pages 392–397, 2012.

R. Krauss and J. Croxell. A low cost microcontroller-in-the-loop platform for control education. In *2012 American Control Conference (ACC)*, pages 4478–4483, 2012.

J. J. Kuffner and S. M. LaValle. RRT-connect: an efficient approach to single-query path planning. In *Proceedings 2000 ICRA. Millennium Conference. IEEE International Conference on Robotics and Automation. Symposia Proceedings (Cat. No. 00CH37065)*, volume 2, pages 995–1001. IEEE, 2000.

T. Kulshreshtha and A. S. Dhar. CORDIC-based high throughput sliding DFT architecture with reduced error-accumulation. *Circuits, Systems, and Signal Processing*, 37(11):5101–5126, 2018.

R. Kümmerle, R. Triebel, P. Pfaff, and W. Burgard. Monte Carlo localization in outdoor terrains using multilevel surface maps. *Journal of Field Robotics*, 25(6–7):346–359, 2008.

Y. Kuwata, G. A. Fiore, J. Teo, E. Frazzoli, and J. P. How. Motion planning for urban driving using RRT. In *2008 IEEE/RSJ International Conference on Intelligent Robots and Systems*, pages 1681–1686. IEEE, 2008.

S. Kwon, K. W. Yang, and S. Park. An effective Kalman filter localization method for mobile robots. In *2006 IEEE/RSJ International Conference on Intelligent Robots and Systems*, pages 1524–1529. IEEE, 2006.

A. V. Le, V. Prabakaran, V. Sivanantham, and R. E. Mohan. Modified A-Star algorithm for efficient coverage path planning in Tetris inspired self-reconfigurable robot with integrated laser sensor. *Sensors*, 18(8):2585, 2018.

M. C. Lee and M. G. Park. Artificial potential field based path planning for mobile robots using a virtual obstacle concept. In *Proceedings 2003 IEEE/ASME International Conference on Advanced Intelligent Mechatronics (AIM 2003)*, volume 2, pages 735–740. IEEE, 2003.

D. Liberzon. *Switching in Systems and Control*, Systems & Control: Foundations & Applications. Birkhauser, Boston, MA., 2003. ISBN 9780817642976.

F. J. Lin, C. K. Chang, and P. K. Huang. FPGA-based adaptive backstepping sliding-mode control for linear induction motor drive. *IEEE Transactions on Power Electronics*, 22:1222–1231, 2007.

M. Lin, J. Yoon, and B. Kim. Self-driving car location estimation based on a particle-aided unscented kalman filter. *Sensors*, 20(9):2544, 2020.

J. Lu and L. J. Brown. A multiple Lyapunov functions approach for stability of switched systems. In *Proceedings of the 2010 American Control Conference*, pages 3253–3256, 2010.

Y. Luo, Y. Wang, Y. Ha, Z. Wang, S. Chen, and H. Pan. Generalized hyperbolic CORDIC and its logarithmic and exponential computation with arbitrary fixed base. *IEEE Transactions on Very Large Scale Integration (VLSI) Systems*, 27(9):2156–2169, 2019.

H. Mahdavi and S. Timarchi. Area-time-power efficient FFT architectures based on binary-signed-digit cordic. *IEEE Transactions on Circuits and Systems I: Regular Papers*, 66(10):3874–3881, 2019.

P. Martí, M. Velasco, J. M. Fuertes, A. Camacho, and G. Buttazzo. Design of an embedded control system laboratory experiment. *IEEE Transactions on Industrial Electronics*, 57(10):3297–3307, 2010.

P. K. Meher and S. Y. Park. Design of cascaded cordic based on precise analysis of critical path. *Electronics*, 8(4), 2019. ISSN 2079-9292.

P. K. Meher, J. Valls, T.-B. Juang, K. Sridharan, and K. Maharatna. 50 years of cordic: algorithms, architectures, and applications. *IEEE Transactions on Circuits and Systems I: Regular Papers*, 56(9):1893–1907, 2009.

J. P. Merlet. Interval analysis for certified numerical solution of problems in robotics. *International Journal of Applied Mathematics and Computer Science*, 2009.

J. P. Merlet. Interval analysis and robotics. In *Robotics Research. Springer Tracts in Advanced Robotics*, pages 147–156. Springer, Berlin, Heidelberg, 2010.

A. Meyer-Bäse, R. Watzel, U. Meyer-Bäse, and S. Foo. A parallel CORDIC architecture dedicated to compute the Gaussian potential function in neural networks. *Engineering Applications of Artificial Intelligence*, 16(7–8):595–605, 2003.

J. Minguez and L. Montano. Nearness diagram navigation (ND): a new real time collision avoidance approach. In *Proceedings. 2000 IEEE/RSJ International Conference on Intelligent Robots and Systems (IROS 2000)(Cat. No. 00CH37113)*, volume 3, pages 2094–2100. IEEE, 2000.

J. Minguez and L. Montano. Nearness diagram (ND) navigation: collision avoidance in troublesome scenarios. *IEEE Transactions on Robotics and Automation*, 20(1):45–59, 2004.

J. Minguez, L. Montano, T. Siméon, and R. Alami. Global nearness diagram navigation (GND). In *Proceedings 2001 ICRA. IEEE International Conference on Robotics and Automation (Cat. No. 01CH37164)*, volume 1, pages 33–39. IEEE, 2001.

M. Moallem. A laboratory testbed for embedded computer control. *IEEE Transactions on Education*, 47(3):340–347, 2004.

C. Moeslinger, T. Schmickl, and K. Crailsheim. A minimalist flocking algorithm for swarm robots. In G. Kampis, I. Karsai, and E. Szathmáry, editors, *Advances in Artificial Life. Darwin Meets von Neumann*, pages 375–382. Springer, Berlin, Heidelberg, 2011. ISBN 978-3-642-21314-4.

R. E. Moore, R. B. Kearfott, and M. J. Cloud. *Introduction to Interval Analysis*. SIAM, 2009.

R. M. Murray and S. S. Sastry. Nonholonomic motion planning: steering using sinusoids. *IEEE Transactions on Automatic Control*, 38(5):700–716, 1993.

S. Murray, W. Floyd-Jones, Y. Qi, D. J. Sorin, and G. D. Konidaris. Robot motion planning on a chip. In *Proceedings of Robotics: Science and Systems*, 2016: AnnArbor, Michigan.

K. Ogata. *Discrete-Time Control Systems*. Prentice-Hall, Inc., USA, 1987. ISBN 0132161028.

P. Ogren and N. E. Leonard. A convergent dynamic window approach to obstacle avoidance. *IEEE Transactions on Robotics*, 21(2):188–195, 2005.

R. Olfati-Saber. Near-identity diffeomorphisms and exponential /spl epsi/-tracking and /spl epsi/-stabilization of first-order nonholonomic SE(2) vehicles. In *ACC: proceedings of the 2002 American Control Conference (8–10 May 2002). Hilton Anchorage and Egan Convention Center, Anchorage, Alaska, USA*, volume 6, pages 4690–4695. 2002.

S. Palnitkar. *Verilog HDL: A Guide to Digital Design and Synthesis*. Prentice-Hall, Inc., USA, 1996. ISBN 0134516753.

A. A. Panchpor, S. Shue, and J. M. Conrad. A survey of methods for mobile robot localization and mapping in dynamic indoor environments. In *2018 Conference on Signal Processing and Communication Engineering Systems (SPACES)*, pages 138–144. IEEE, 2018.

K. K. Parhi. *VLSI Digital Signal Processing Systems: Design and Implementation*. John Wiley & Sons Inc., 2007. ISBN 9780471241867.

C. A. C. Parker and H. Zhang. Consensus-based task sequencing in decentralized multiple-robot systems using local communication. In *2008 IEEE/RSJ International Conference on Intelligent Robots and Systems*, pages 1421–1426, 2008.

D. Pickem, M. Lee, and M. Egerstedt. The gritsbot in its natural habitat - a multi-robot testbed. In *2015 IEEE International Conference on Robotics and Automation (ICRA)*, pages 4062–4067, 2015.

B. Plancher, S. M. Neuman, T. Bourgeat, S. Kuindersma, S. Devadas, and V. J. Reddi. Accelerating robot dynamics gradients on a CPU, GPU, and FPGA. *IEEE Robotics and Automation Letters*, 6(2):2335–2342, 2021.

S. Quinlan and O. Khatib. Elastic bands: connecting path planning and control. In *1993 Proceedings IEEE International Conference on Robotics and Automation*, pages 802–807. IEEE, 1993.

C. A. Rabbath and N. Léchevin. *Discrete-Time Control System Design with Applications*. Springer Publishing Company, Incorporated, 2014. ISBN 9781461492894.

Y. Rasekhipour, A. Khajepour, S. K. Chen, and B. Litkouhi. A potential field-based model predictive path-planning controller for autonomous road vehicles. *IEEE Transactions on Intelligent Transportation Systems*, 18(5):1255–1267, 2016.

M. Rubenstein, C. Ahler, and R. Nagpal. Kilobot: A low cost scalable robot system for collective behaviors. In *2012 IEEE International Conference on Robotics and Automation*, pages 3293–3298, 2012.

M. G. Ruppert, D. M. Harcombe, and S. O. R. Moheimani. High-bandwidth demodulation in MF-AFM: a Kalman filtering approach. *IEEE/ASME Transactions on Mechatronics*, 21(6):2705–2715, 2016.

A. Sabanovic, L. M. Fridman, and S. K. Spurgeon. *Variable Structure Systems: From Principles to Implementation*. Control, Robotics & Sensors. Institution of Engineering and Technology, 2004.

E. Sahin. Swarm Robotics: from source to inspiration to domains of application. In *Int'l Workshop on Swarm Robotics*, 2004. ISBN 978-3-540-24296-3. https://doi.org/10.1007/b105069.

O. Salzman and D. Halperin. Asymptotically near-optimal rrt for fast, high-quality motion planning. *IEEE Transactions on Robotics*, 32(3):473–483, 2016.

M. Seder and I. Petrovic. Dynamic window based approach to mobile robot motion control in the presence of moving obstacles. In *In Proceedings 2007 IEEE International Conference on Robotics and Automation*, pages 1986–1991, 2007.

W. J. Seo, S. H. Ok, J. Ahn, S. Kang, and B. Moon. An efficient hardware architecture of the A-Star algorithm for the shortest path search engine. In *2009 5th International Joint Conference on INC, IMS and IDC*, pages 1499–1502. IEEE, 2009.

D. Shah and L. Vachhani. Swarm aggregation without communication and global positioning. *Joint Publication in IEEE Robotics and Automation Letters*, 4(2):886–893, 2019 *and in Proceedings of International Conferenceon Robotics and Automation (ICRA) 2019, Montreal,Canada.*, 2019.

Y. Shtessel, C. Edwards, L. Fridman, and A. Levant. *Sliding Mode Control and Observation*. Control Engineering. Birkhäuser Basel, 1st edition, 2014. ISBN 9780817648923.

R. Shukla and K. C. Ray. Low latency hybrid cordic algorithm. *IEEE Transactions on Computers*, 63(12):3066–3078, 2014.

B. Siciliano and O. Khatib. *Springer Handbook of Robotics*. Springer International Publishing, 2016. ISBN 9783319325507.

R. Siegwart, I. R. Nourbakhsh, and D. Scaramuzza. *Introduction to Autonomous Mobile Robots*. MIT Press, 2011. ISBN 9780262015356.

R. Simmons. The curvature-velocity method for local obstacle avoidance. In *Proceedings of IEEE International Conference on Robotics and Automation*, pages 3375–3382, 1996.

C. J. Sun, H. Y. Kuo, and C. E Lin. A sensor based indoor mobile localization and navigation using unscented Kalman filter. In *IEEE/ION Position, Location and Navigation Symposium*, pages 327–331. IEEE, 2010.

I. Susnea, A. Filipescu, G. Vasiliu, G. Coman, and A. Radaschin. The bubble rebound obstacle avoidance algorithm for mobile robots. In *IEEE ICCA 2010*, pages 540–545. IEEE, 2010.

G. Tan, H. He, and S. Aaron. Global optimal path planning for mobile robot based on improved Dijkstra algorithm and ant system algorithm. *Journal of Central South University of Technology*, 13(1):80–86, 2006.

S. Thrun, D. Fox, W. Burgard, and F. Dellaert. Robust Monte Carlo localization for mobile robots. *Artificial Intelligence*, 128(1–2):99–141, 2001.

S. Thrun, W. Burgard, and D. Fox. *Probabilistic Robotics, Intelligent Robotics and Autonomous Agents*. The MIT Press, 2005. ISBN 9780262201629.

C. S. Tzafestas and S. G. Tzafestas. Recent algorithms for fuzzy and neurofuzzy path planning and navigation of autonomous mobile robots. *Systems Science*, 25(2):25–39, 1999.

I. Ullah, Y. Shen, X. Su, C. Esposito, and C. Choi. A localization based on unscented Kalman filter and particle filter localization algorithms. *IEEE Access*, 8:2233–2246, 2019.

C. Urmson and R. Simmons. Approaches for heuristically biasing RRT growth. In *Proceedings 2003 IEEE/RSJ International Conference on Intelligent Robots and Systems (IROS 2003)(Cat. No. 03CH37453)*, volume 2, pages 1178–1183. IEEE, 2003.

L. Vachhani. Education on architecture development for embedded controllers. *IFAC Proceedings Volumes*, 46(17), 61–65, 2013.

L. Vachhani. CORDIC as a switched nonlinear system. *Circuits System and Signal Processing*, 2020 39, 3234–3249, 2020. https://doi.org/10.1007/s00034-019-01295-8.

L. Vachhani, K. Sridharan, and P.K. Meher. Efficient CORDIC algorithms and architectures for low area and high throughput implementation. *IEEE Transactions on Circuits and Systems II: Express Briefs*, 56(1):61–65, 2009.

F. Vahid and R. Lysecky. *VHDL for Digital Design*. John Wiley & Sons, 2007. ISBN 9780470052631.

J. E. Volder. The CORDIC trigonometric computing technique. *IRE Transactions on Electronic Computers*, EC-8(3):330–334, 1959.

P. Vyas. Interval arithmetic methods for hardware-efficient implementation of high-level robotic tasks. PhD thesis. Indian Institute of Technology Bombay, India, 2017.

P. Vyas, L. Vachhani, K. Sridharan, and V. Pudi. CORDIC-based azimuth calculation and obstacle tracing via optimal sensor placement on a mobile robot. *IEEE/ASME Transactions on Mechatronics*, 21(5):2317–2329, 2016.

P. Vyas, L. Vachhani, and K. Sridharan. Hardware-efficient interval analysis based collision detection and avoidance for mobile robots. *Mechatronics*, 62, 2019.

P. Vyas, L. Vachhani, and K Sridharan. Interval analysis technique for versatile and parallel multi-agent collision detection and avoidance. *Journal of Intelligent and Robotic Systems*, 98(3):705–720, 2020.

T. J. Vyncke, S. Thielemans, and J. A. Melkebeek. Finite-set model-based predictive control for flying-capacitor converters: cost function design and efficient FPGA implementation. *IEEE Transactions on Industrial Informatics*, 9(2):1113–1121, 2013.

E. Walter, L. Jaulin, and M. Kieffer. Interval analysis for guaranteed and robust nonlinear estimation in robotics. In *3rd World Congress of Nonlinear Analysts (WCNA)*, page 191–202, 2001.

Z. Wan, B. Yu, T. Y. Li, J. Tang, Y. Zhu, Y. Wang, A. Raychowdhury, and S. Liu. A survey of FPGA-based robotic computing. *IEEE Circuits and Systems Magazine*, 21(2):48–74, 2021.

Y. Wang and G. S. Chirikjian. A new potential field method for robot path planning. In *Proceedings 2000 ICRA. Millennium Conference. IEEE International Conference on Robotics and Automation. Symposia Proceedings (Cat. No. 00CH37065)*, volume 2, pages 977–982. IEEE, 2000.

H. Wang, Y. Yu, and Q. Yuan. Application of Dijkstra algorithm in robot path-planning. In *2011 2nd International Conference on Mechanic Automation and Control Engineering*, pages 1067–1069. IEEE, 2011.

T. Wescott. *Applied Control Theory for Embedded Systems*. Embedded Technology. Newnes, Burlington, 2006. ISBN 9780750678391.

R. L. Williams and D. A. Lawrence. *Linear State-Space Control Systems*. John Wiley & Sons, Inc., 2007. ISBN 9780471735557.

A. G. Wills, G. Knagge, and B. Ninness. Fast linear model predictive control via custom integrated circuit architecture. *IEEE Transactions on Control Systems Technology*, 20(1):59–71, 2012.

W. M. Wonham. Linear multivariable control. *Optimal Control Theory and its Applications*, pages 392–424, Springer, 1974.

F. Yao, Q. Zhou, and Z. Wei. A novel multilevel RF-PWM method with active-harmonic elimination for all-digital transmitters. *IEEE Transactions on Microwave Theory and Techniques*, 66(7):3360–3373, 2018.

K. N. Yong and R. G. Simmons. The lane-curvature method for local obstacle avoidance. In *In Proceedings. 1998 IEEE/RSJ International Conference on Intelligent Robots and Systems*, pages 1615–1621, 1998.

Z. Zhang and Z. Zhao. A multiple mobile robots path planning algorithm based on A-Star and Dijkstra algorithm. *International Journal of Smart Home*, 8(3):75–86, 2014.

Index

Embedded Control for Mobile Robotic Applications, First Edition.
Leena Vachhani, Pranjal Vyas, and Arunkumar G. K.
© 2022 The Institute of Electrical and Electronics Engineers, Inc. Published 2022 by John Wiley & Sons, Inc.
Companion website: www.wiley.com/go/vachhani/embeddedcontrolforroboticapp

Books in the IEEE Press Series on Control Systems Theory and Applications

Series Editor: Maria Domenica DiBenedetto, University of l'Aquila, Italy

The series publishes monographs, edited volumes, and textbooks which are geared for control scientists and engineers, as well as those working in various areas of applied mathematics such as optimization, game theory, and operations.

1. *Autonomous Road Vehicle Path Planning and Tracking Control*
   Levent Güvenç, Bilin Aksun-Güvenç, Sheng Zhu, Şükrü Yaren Gelbal

2. *Embedded Control for Mobile Robotic Applications*
   Leena Vachhani, Pranjal Vyas, and Arunkumar G. K.